Problem Solving
Book A

MATH TO KNOW

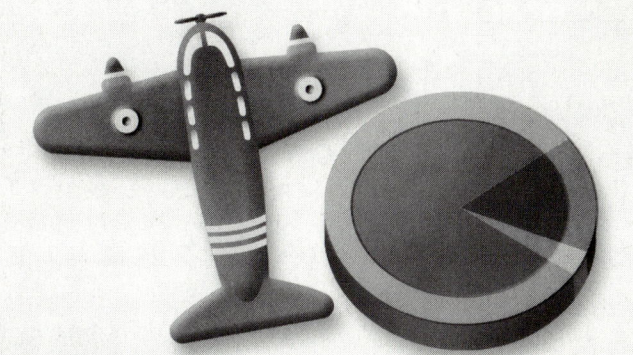

GReaT SouRCe
EDUCATION GROUP
A Houghton Mifflin Company
New Ways to Know

Developing thinking skills through **READING and WRITING**

Acknowledgments and Credits

Acknowledgments
Reviewers:

Cathy Bernhard
    Mathematics Consultant
    Portland, OR

Marilyn R. LeRud
    Retired Elementary Education Teacher
    Tucson Unified School District
    Tucson, AZ

Beth Spivey
    Coordinating Teacher
    Elementary Mathematics
    Wake County Public Schools
    Raleigh, NC

Credits

Writing: Justine Dunn, Judy Vandegrift

Review: Edward Manfre

Editorial: Carol DeBold, Pearl Ling, Susan Rogalski

Design/Production: Taurins Design

Creative Art: Joe Boddy *pages 90, 92, 105*. Alex Farquharson *pages 10, 19, 50, 51, 73, 74, 75, 83 (middle), 84, 85, 104*. Greg Harris *pages 20, 78, 91*. Amanda Harvey *pages 14, 18, 29, 33, 39, 46*. Eileen Hine *page iv, icons*. Loretta Lustig *pages 11, 12, 17, 27, 28, 37, 38, 40, 47, 58, 68, 69, 80, 81, 82, 83 (top & bottom), 87*. Steve Mach *pages 2-3, 4-5, 24, 36, 49, 59, 61, 63*.

Technical Art: Taurins Design

Photos: Corbis *pages iv, 1, 8, 9, 18 (background & insets 1, 2, & 4), 19, 24, 26, 30, 31, 36, 37 (top), 44, 54, 55 (top & middle), 56, 64, 65, 66, 67, 69, 72, 73, 79, 86, 90, 91, 94, 97, 98 (top right), 99, 100, 101*. Franklin County Arts Council *page 55 (bottom)*. NASA *pages 6, 7, 41*. National Geographic Image Collection *page 98 (top left)*. Novica.com *pages 18 (inset 3), 21*. Okanagan University College *page 59*. Rube Goldberg *page 37 (bottom)*. Sandia National Laboratories *page 42*. Tournament of Roses/Long Photography *page 19 (bottom right)*. White House Historical Association *page 83*. John Zweifel *pages 73 (bottom left & right), 75, 77, 80, 81, 87*.

Cover Design: Kristen Davis

Copyright © 2005 by Great Source Education Group, a division of Houghton Mifflin Company. All rights reserved.

No part of this work may be reproduced or transmitted in any form or by any means electronic or mechanical, including photocopying and recording, or by any information storage retrieval system without the prior written permission of Great Source Education Group, unless such copying is expressly permitted by federal copyright law. Address inquires to Permissions, Great Source Education Group, 181 Ballardvale Street, Wilmington, MA 01887.

Great Source® is a registered trademark of Houghton Mifflin Company.

All registered trademarks are shown strictly for illustrative purposes and are the property of their respective owners.

Printed in the United States of America.

International Standard Book Number: 0–669–50047–X

3 4 5 6 7 8 9 –POO– 10 09 08 07 06 05

Visit our web site: http://www.greatsource.com

Table of Contents

Chapter 1 — Groups Big and Small
Reading Piece by Piece 1
   Understanding Common Words or Phrases 2
   Visualizing a Math Sentence 10
   Using Math Symbols 14
   Chapter Test 16

Chapter 2 — Amazing Animals
Finding Exactly What You Need 18
   Finding Information You Need to Solve a Problem ... 20
   Problems with Missing Information 30
   Chapter Test 34

Chapter 3 — Creative Inventions
Making a Plan 36
   Visualizing a Math Problem 38
   Making a Plan 46
   Chapter Test 52

Chapter 4 — Contests and Races
Carrying Out the Plan 54
   Choosing the Correct Solution 56
   Using a Plan to Solve a Word Problem 62
   Writing the Plan and Solving the Word Problem 66
   Chapter Test 70

Chapter 5 — Mighty Castles
Looking Back 72
   Answering the Question Asked 74
   Comparing Your Answer to Another Number 78
   Using the Correct Label for Your Answer 82
   Using Estimation to Check Your Answer 84
   Chapter Test 88

Chapter 6 — Ribbit, Ribbit
Putting It All Together 90
   Using the Four-Step Problem-Solving Method 92
   Solving Math Problems on Your Own 102
   Chapter Test 106

Vocabulary 108

Chapter 1

Groups Big and Small

Reading Piece by Piece

Calling All Artists!

The public library is making a Celebrate Communities display. We need more pictures of groups big and small to add to the mural. All types of communities are needed— for example, birds, fish, or people.

Draw a picture of your favorite community.
Send it in!

▶ **Understand**   Plan   Try   Look Back

In this chapter, you will learn about different communities. When you live in a community, you need to be able to speak with others and understand what others are saying to you. Sometimes, people use special words. In this chapter, you will learn about some special math words and symbols. You'll get lots of practice using the first step of the four-step problem solving method: **Understand**.

There are many different kinds of communities. You live in a community. You might live in a big city or in a very small town.

▲ Everyday you go to school. Schools are communities made of students, teachers, and staff.

Some animals live in communities.

▼ Whales live in groups called *pods*.

▲ Geese live in groups called *flocks*.

1

Understanding Common Words or Phrases

Sometimes the same word can mean different things.

The math meaning of a word may be different from the everyday meaning of the same word.

Think about places in your community; for example, the playground.

PLAYHOUSE
for children under 10 years old

Look for the vocabulary word on this page. Circle it, then go to the Vocabulary Section, which begins on page 108. Write a definition for the word. Include diagrams or examples.

Vocabulary ▾ under

2

▶ **Understand**   Plan   Try   Look Back

Sometimes one word, such as *under*, can have different meanings. See how the word is used to get an idea of its meaning.

Look at the picture of the playground.

The sign says you must be <u>under</u> 10 years old to go in the playhouse.

The backpack is <u>under</u> the sign.

**Look at the picture.
Circle one word in the box that will make both sentences true.
Write that word in each sentence.**

| eyes | (swing) | feet |

1. The girl has 2 __swings__.

2. The swing is 2 _____ high. ◀ MTK 346

This number tells you where to find more information in *Math to Know*.

more ▶

Vocabulary ▪ feet (ft)

3

Sometimes a math word can have more than one math meaning.

Think of the word *even*. It can mean a number such as 2, 4, or 6. The boy wearing the cap has an *even* number on his shirt. It can also mean having the same number in each group. There are 3 players on each team. So, the teams are *even*.

Look at the picture.
Circle one word in the box that will make both sentences true.
Write that word in each sentence.

| right | left | boy |

3. The girl with a 3 on her shirt is pointing to the _____.

4. Two birds flew away. There are 3 birds _____. ◀ MTK 46

▶Understand   Plan   Try   Look Back

The word *third* can mean a fraction. When something is divided into 3 equal parts, each section is one *third*. The word *third* can also mean what place something is in.

Look at the picture. Write two sentences about the picture. Use the word *third* in a different way in each sentence. ◀MTK 210, 16

5. _____

6. _____

more ▶

Vocabulary ▼ third ▼ one third ▼ fraction   5

Sometimes different words can mean the same thing.

Some communities are in outer space. Often groups of astronauts live and work in a space station.

Astronauts need to learn how to live when they are away from Earth's gravity. They train on a KC-135 jet that takes such fast dives that many of them become ill while in flight. Because of this, the KC-135 is also called the *Vomit Comet*.

In math, we also have different ways to name the same thing.

Circle all the words or numbers that mean the same as the underlined part of the sentence. Circle all that apply.

7. An astronaut living in space needs <u>at least 2</u> pounds of oxygen a day. ◀ MTK 12

   less than 2   2 or more   2 or less

8. The record length of time for someone to live in a space station is <u>436</u> days. ◀ MTK 5

   four hundred thirty-six   4 hundreds, 3 tens, 6 ones   40036

Vocabulary ▼ at least ▼ pound (lb) ▼ less than (<)

▶ **Understand** Plan Try Look Back

9. Astronauts need <u>one third</u> of a day for sleep. ◀ MTK 210-211

   three   13 $\frac{1}{3}$

10. The Russian space station *Mir* has traveled <u>more than 1 billion</u> miles while circling Earth. ◀ MTK 11, 12

    Did you know?
    The Russian space station is named *Mir*, which means *peace*.

    over 1 billion   greater than 1 billion   less than 1 billion

Write another word or phrase that means the same as the underlined part of the sentence.

11. The designers of *Skylab* had <u>201</u> T-shirts put in the space station for the crew. ◀ MTK 5

    Did you know?
    Skylab was the first American space station.

    _____

    _____

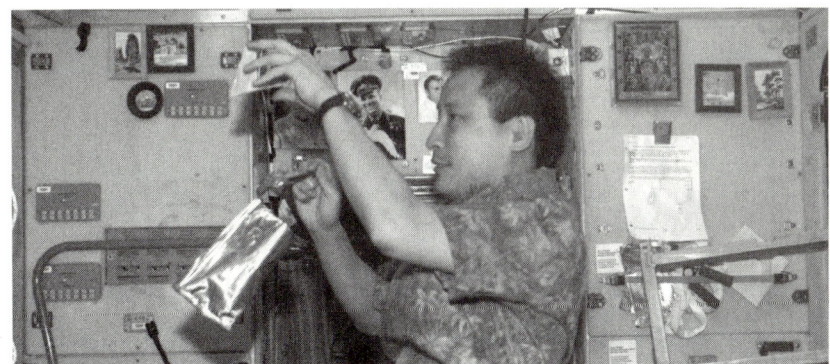

12. The movie *Star Wars* has a space station in it. It costs <u>$4</u> to rent the *Star Wars* DVD. ◀ MTK 17

    _____

    _____

    more ▶

Vocabulary ▬ billion ▬ mile (mi) ▬ greater than (>) ▬ dollar ($)

Some numbers tell exact amounts. Other numbers just tell *about* how many or *about* how much.

Look at the sentence below.

Each ant has 6 legs.

In this sentence, the number 6 is an exact number.

Now look at this sentence.

There are over 9,000 different kinds of ants.

The number 9,000 is a rounded number. It gives an estimate of the number of kinds of ants.

Sometimes you need an exact number and sometimes an estimate is just fine.

Read the story below. Underline the numbers you think are exact. Circle the numbers you think are (estimates). ◀ MTK128–130

13.

 **Ants**

Ants live in communities. Their communities are called *colonies*.

Some ants build colonies that reach about (20) feet below the ground. One group of ants built a nest that was exactly 5 feet above the ground.

An ant has 2 stomachs and 2 eyes. Most ants live around 50 days. The queen ant may live for about 20 years.

Ants are very strong. A worker ant can lift about 50 times its body weight.

Did you know?
Some ants can live up to 14 days underwater.

8   Vocabulary ▾ exact ▾ estimate ▾ weight

Read the story below. Underline the numbers you think are exact. Circle the numbers you think are (estimates). ◀ MTK128–130

14.

Bees

Bees live in communities, too. They build hives. Each cell of the hive is in the shape of a hexagon. Each cell has 6 sides.

A hive has one queen bee, drone bees, and worker bees. Worker bees live for about 30 days. The queen bee lives about 4 years.

Honeybees usually fly about 2 miles from their hive to look for flowers. They can visit almost 1,000 flowers a day. They get nectar from flowers to make honey. It takes 4 pounds of nectar to make 1 pound of honey.

Bees have 2 large eyes and 3 small eyes. They have 6 legs and 4 wings.

Do you know why bees buzz? It is because they move their wings about 11,000 times in 1 minute. The buzz is the noise from their wings flapping so quickly.

Did you know? The honeybee is the state insect of 10 states.

Vocabulary ▬ hexagon ▬ side ▬ minute (min)

Visualizing a Math Sentence

Sometimes it helps to make a picture of what a sentence tells you.

To help you understand a math sentence, make a picture of what you see in your mind when you read the sentence.

What do you see when you read this sentence?

There are 3 flowers with 2 bees on each flower.

Here are two different ways you can picture the sentence.

You can draw a very realistic picture with lots of details. Or, you can draw a simple sketch to show the math.

Realistic Picture

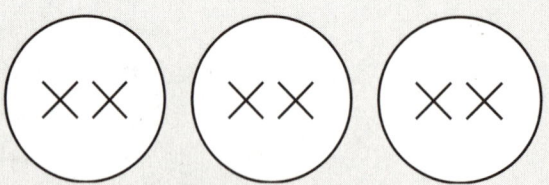

Simple Sketch

A simple sketch can help you pay attention to what's important. It also takes less time to draw.

▶ **Understand**   Plan   Try   Look Back

In front of exercises 1–3, write the letter of the picture that best shows the meaning of the math.

_____ 1. The bee flew 3 miles of its 5-mile journey to the flowerbed. ◀ MTK 239

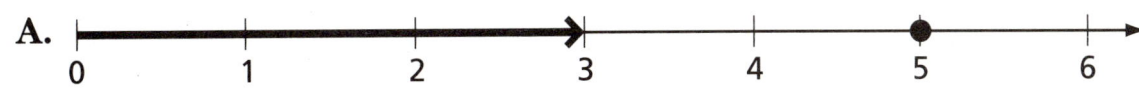

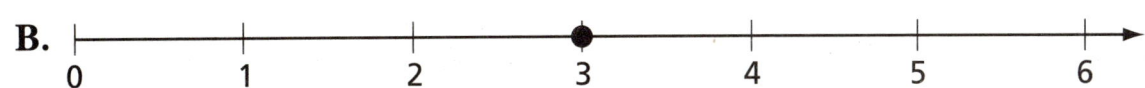

_____ 2. There are 4 rooms underground. Three ants are in each room.

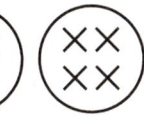 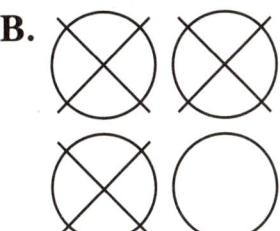 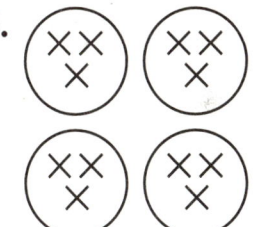

_____ 3. Some large anthills can be 3 feet tall. ◀ MTK 346

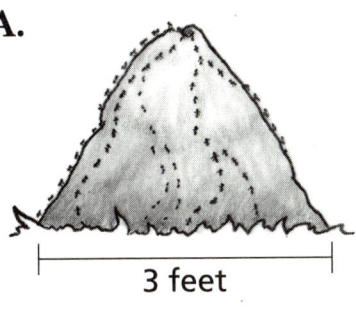   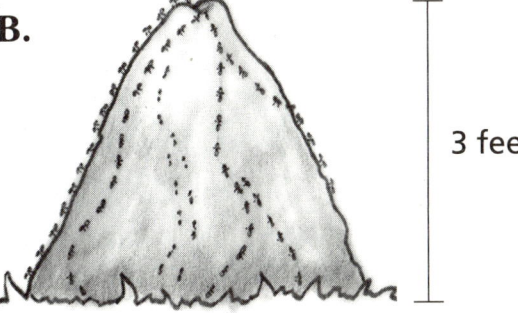

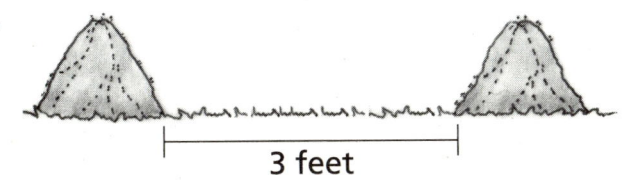

more ▶

11

Useful math pictures show what is needed to solve a word problem.

In front of exercises 4–6, write the letter of the picture that best shows the meaning of the math.

_____ 4. When the first crew entered the *Skylab* space station, the temperature was 130 degrees Fahrenheit. ◀ MTK 360

A.
B.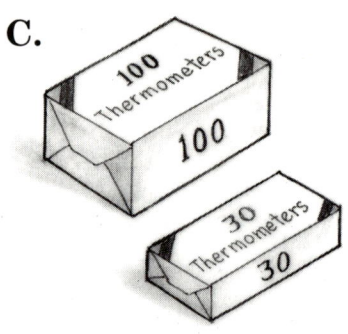
C.

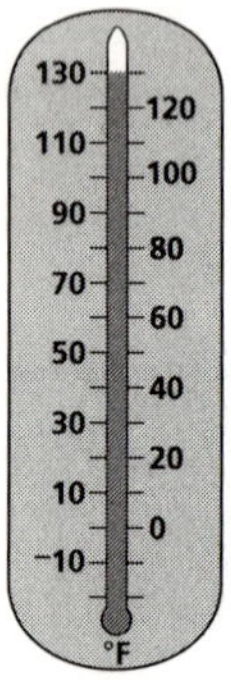

_____ 5. The *Skylab* astronauts lived and worked in a rectangular room that was 48 feet long and 22 feet wide. ◀ MTK 312–313, 346

A.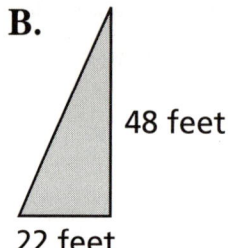
B.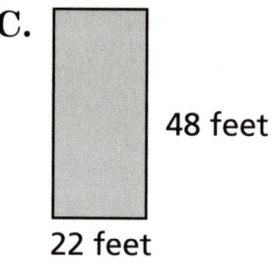
C.

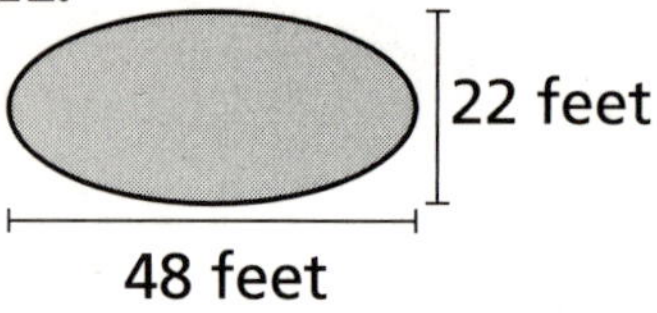

_____ 6. The crew brought 2 spiders and 6 mice onto the *Skylab* space station.

A.
B.
C.

12   Vocabulary ▼ temperature ▼ degrees (°) ▼ Fahrenheit (F) ▼ rectangular

> **Understand** Plan Try Look Back

Draw a picture to represent the sentence.

7. Carter was fifth in line at the cafeteria. ◀ MTK 16

8. The large laboratory on *Skylab* was almost 50 meters long by 7 meters wide. ◀ MTK 347

9. The competition is between 2 teams of 7 players each.

10. Three of the ants are climbing up the anthill, while 5 are climbing down.

Vocabulary ▼ meter (m) ▼ fifth

13

Using Math Symbols

Symbols can stand for one word or a group of words.

Honeybees do a waggle dance to tell other bees where flowers are located. They dance in a figure-eight pattern. Here are some symbols.

Go to the left to get to the flowers.

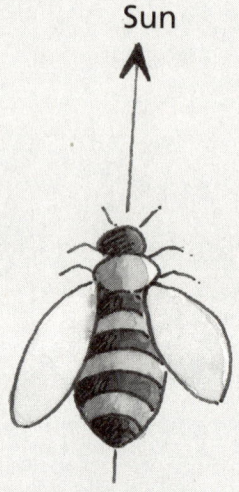

Go straight ahead to get to the flowers.

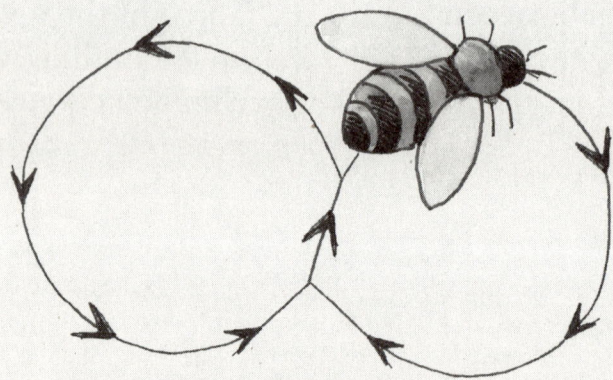

Go to the right to get to the flowers.

In math, we often use symbols as a shortcut to write a word or group of words.

Sometimes when you are solving a math problem, it helps if you can replace some of the words with a symbol.

14

> **Understand** Plan Try Look Back

Write one of the symbols from below in the ◯ to compare the information. ◀ MTK 12, 471–472

$$+ \quad - \quad \times \quad \div \quad =$$

$>$ *is greater than*　　$<$ *is less than*

1. There were 3 bumblebees buzzing in one direction and 3 bumblebees buzzing in the other direction.

   The bumblebees buzzing　◯　The bumblebees buzzing
   in one direction　　　　　　in the other direction

2. There were 5 fire ants and 6 carpenter ants on the back porch.

   The number of fire ants ◯ The number of carpenter ants

3. A hexagon has 6 sides and a square has 4 sides. ◀ MTK 311–313

   The number of sides　◯　The number of sides
   on a hexagon is　　　　　on a square

4. The number of miles from the hive to the flower garden and back to the hive is a total of 8 miles.

   The number of miles from　◯　The number of miles from
   the hive to the garden　　　　the garden to the hive

You have learned a lot about different types of communities. You are now ready to draw a picture of your favorite community.

Vocabulary ▼ square

Chapter 1 Test

Fill in the circle with the letter of the correct answer.

1. Name the third shape from the left.

     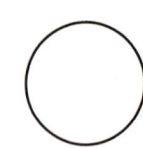 

   (A) triangle (B) circle (C) square

2. Which circle is divided into thirds?

   A 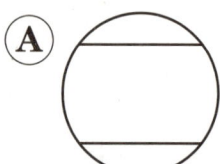 B 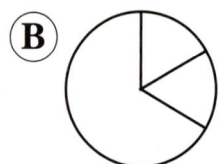 C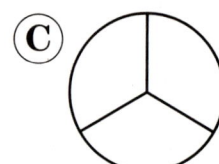

3. The video costs <u>less than 5</u> dollars.

   Which group of words means the same as the underlined words?

   (A) under 5 (B) exactly 5 (C) more than 5

4. There were about 50 people in the park. I saw 4 people on rollerblades and a boy carrying 2 tennis rackets.

   In the sentences above, which number is an estimate?

   (A) 2 (B) 50 (C) 4

5. There are more laptop computers than desktop computers.

   Which statement gives the same information as the sentence above?

   (A) The number of laptop computers < The number of desktop computers

   (B) The number of laptop computers > The number of desktop computers

   (C) The number of desktop computers > The number of laptop computers.

16

Choose the letter of the best answer. Then write why you made that choice.

6. Suki has 2 dogs and 3 fish.

   Choose the picture that best shows the meaning of the sentence above.

   A) 🐕 🐕 🐟 🐟 🐟 _____

   B) 🐕 🐕 🐕 🐟 🐟 _____

   C) 🐕 🐟 _____

Answer the questions.

7. Explain, using words or simple sketches, two math meanings for the word *fifth*.

   • _____

   • _____

8. Draw a simple sketch to show the math in this sentence.

   There are 2 boxes with 3 pencils in each box.

   Explain why your sketch shows the math in the sentence.

   _____   **Draw your sketch here.**

   _____

   _____

   _____

17

Chapter 2

Amazing Animals
Finding Exactly What You Need

TV Notes
The Animal Quiz Show
Contest Theme: Amazing Animals

HELP WANTED

Contestants Wanted for *The Animal Quiz Show.* The theme is Amazing Animals. First prize winners choose from the following fabulous trips:

- Visit *T. rex Sue* in Chicago
- Whale watch in the Pacific
- Paint with elephant artists in Thailand
- Have tea with Koko the Gorilla

Understand Plan Try Look Back

In this chapter, you are a contestant getting ready for *The Animal Quiz Show* on television. You will come across all kinds of interesting data in paragraphs, charts, tables, and graphs. You will also learn how to find just the information needed to solve a math word problem. This will help you with the first step of the four-step problem-solving method: **Understand**.

◀ Elephants are the largest animals that live on land. Their trunks are strong enough to lift you and one of your friends at the same time.

◀ Gorillas live in Africa. Each group has between 3 to 30 gorillas.

▼ Even though whales live in the water, they are not fish. They live in the water, but they breathe air just like other land animals.

▼ Dinosaurs lived on Earth for 150 million years. We learn a lot about them by looking at fossils, which are rocks that show what their bones looked like.

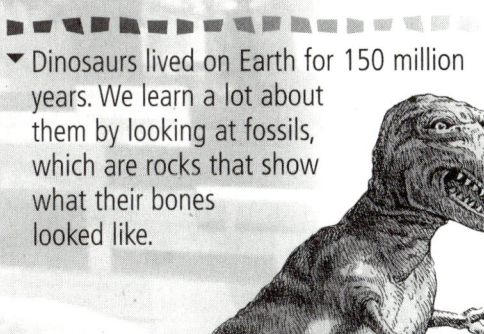

Get set to learn about some very amazing animals. Discover the world of artistic elephants, a dinosaur from long ago, a whale that got lost, and a talking gorilla.

19

Finding Information You Need To Solve a Problem

Information may be found in paragraphs, charts, or graphs.

You think it will be a lot of fun to be a contestant on *The Animal Quiz Show*. So, you sign up. Now you have to learn as much as you can about famous animals. To get started, you and some friends write a few questions.

You will find the answers to these questions and many others as you learn about the interesting things some animals can do. You might have to read very carefully so that you can find the exact information you need. Remember, finding the right information will help you solve math word problems, too.

20   Vocabulary ▾ graph

▶ **Understand** Plan Try Look Back

You have heard about elephants that paint. You decide to read more about them.

In 1998, 2 Russian artists started an art school for elephants in Thailand. The Novica Gallery now has paintings from 15 different elephants on their website. There were over 100 elephant paintings on the website. Most of the paintings are about 30 inches long and 22 inches wide. A painting usually costs between $200 and $400.

**Read the paragraph. Then answer exercises 1–5.
Exercise 1 has a hint .**

1. How many Russian artists started the art school for elephants?

   _____

2. How many different elephants have paintings on the Novica website?

   _____

3. How many different elephant paintings were on the Novica website?

   _____

4. What is the usual length of an elephant painting? ◀ MTK 346

   _____

5. An elephant painting usually costs between what amounts of money? ◀ MTK 17

   _____

more ▶

Vocabulary ▼ inch (in.) ▼ long ▼ wide

21

Pictographs and charts can show lots of information.

You look at the elephant art website to check out the paintings. You make a pictograph to show the number of pieces painted by some of the elephant artists.

Elephant Art

Name of Elephant	Number of Paintings
Arum	🎨🎨🎨🎨🎨🎨🎨🎨
Desi	🎨🎨🎨🎨🎨🎨
Eva	🎨🎨🎨🎨
Jojo	🎨🎨🎨🎨
Tao	🎨

Key: 🎨 = 1 painting

Use the pictograph to answer exercises 6–11. ◀ MTK 270-271

6. How many paintings does each 🎨 stand for? _____

7. How many different elephants are listed on the graph? _____

8. Which two elephants have painted an equal number of paintings?

   _____

9. Which elephant painted the fewest paintings? _____

10. Which elephant painted the second most paintings? _____

11. What is the title of this pictograph? _____

22   Vocabulary ▼ pictograph ▼ equal ▼ second (adj)

> **Understand** Plan Try Look Back

Which elephant paintings do your friends like the best? You decide to take a survey. You use tally marks to record the results.

Favorite Elephant Art

Name of Painting	Votes	
	Tally	Number
Dreams Come True	||||| /	6
I Know I Can	||||| ///	
Just For My Mom	||||| /	
Mars	///	
Violet Trunk	|||||	

Use the information in the chart to answer exercise 12.

12. Complete the **Number** column in the chart.

Use the completed chart to answer exercises 13–17. ◀ MTK 267

13. How many elephant paintings are on the chart? _____

14. Which elephant painting was the favorite? _____

15. Which elephant painting was the least favorite? _____

16. Which painting had the same number of tally marks as *Just For My Mom*?

    _____

17. What is the title of this chart? _____

more ▶

Vocabulary ▼ survey ▼ tally mark

23

Tables can also show information.

In 1990, Susan Hendrickson found some dinosaur bones in South Dakota. These bones belonged to a *Tyrannosaurus rex*, or *T. rex*. When all the bones were collected and pieced together, everyone realized that this was the largest and most complete set of *T. rex* bones ever found. The dinosaur was named *Sue* in honor of Susan Hendrickson. You can visit *Sue* at *The Field Museum* in Chicago.

You learn that *Sue* weighed about 7 tons. A typical *T. rex* weighed about 6 tons. You want to find out if other dinosaurs weighed more than or less than a typical *T. rex*. You do some research and make a table to show the typical estimated weight of a few other dinosaurs.

Sue Hendrickson at the *T. rex Sue* exhibit.

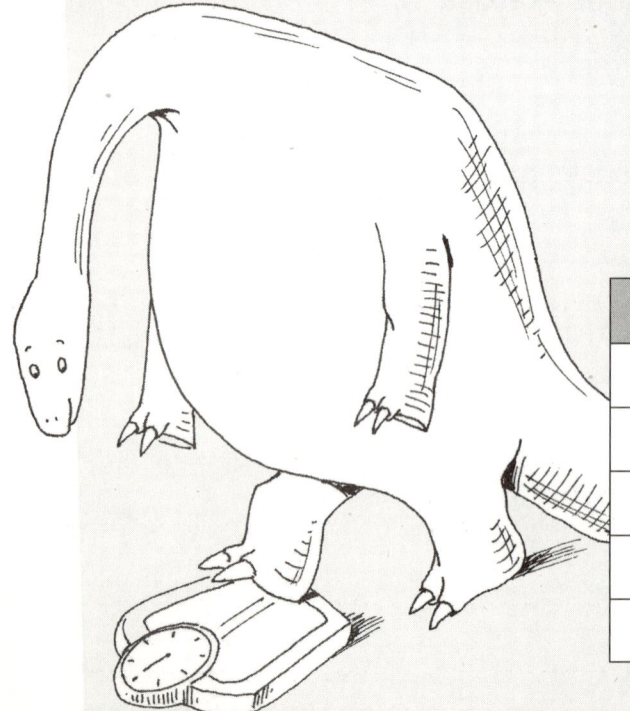

Typical Weights of Dinosaurs

Dinosaur	Weight
Brontosaurus	35 tons
Lambeosaurus	7 tons
Stegosaurus	2 tons
T. rex	6 tons
Triceratops	6 tons

Vocabulary ▼ ton (t) ▼ more than (>) ▼ table

▶ **Understand**   Plan   Try   Look Back

Use the table on page 24 to circle the letter of the correct answer for exercises 18–22. ◀ MTK 268

Did you know?
T. rex Sue lived about 67 million years ago. Scientists think that *T. rex* Sue was an old dinosaur because the bones show signs of wear and tear.

18. Which dinosaur in the table weighed more than a *T. rex*?

   A. Triceratops   **B.** Brontosaurus   **C.** Stegosaurus

19. How many dinosaurs in the table weighed less than a *T. rex*?

   A. 2   **B.** 1   **C.** 3

20. Which dinosaur in the table weighed less than 5 tons?

   A. Brontosaurus   **B.** T. rex   **C.** Stegosaurus

21. Which dinosaur in the table weighed closest to 30 tons?

   A. Brontosaurus   **B.** Lambeosaurus   **C.** Triceratops

22. Which listing below shows the dinosaur weights from heaviest to lightest?

   A. Brontosaurus
      Lambeosaurus
      Stegosaurus
      T. rex or Triceratops

   B. Stegosaurus
      Lambeosaurus
      T. rex or Triceratops
      Brontosaurus

   C. Brontosaurus
      Lambeosaurus
      T. rex or Triceratops
      Stegosaurus

Use the table on page 24 to answer exercise 23. ◀ MTK 268

23. Use the data in the table to write a multiple-choice question. Then ask a friend to answer it.

   _____

   _____

   A. _____   **B.** _____   **C.** _____

more ▶

Vocabulary ▼ million ▼ year ▼ data

25

Bar graphs are a good way to display information.

A friend just told you about a Humpback whale named Humphrey. You want to find out more about him. At the library, you dig up an old newspaper article.

Humphrey, the Wrong-Way-Whale Finds the Right Way

On October 11, 1985, the whale that became known as Humphrey entered San Francisco Bay by accident. Humphrey, a 40-foot Humpback whale, swam 60 miles up the Sacramento River. But whales don't survive in rivers. Lots of people tried to help Humphrey find his way back to the ocean. Finally, on November 4, Humphrey returned to the Pacific Ocean.

What an amazing story! A 40-foot long whale is no goldfish. Imagine one cruising up a river near your hometown. Was Humphrey a big whale? You do some research and make a bar graph to show what you've learned.

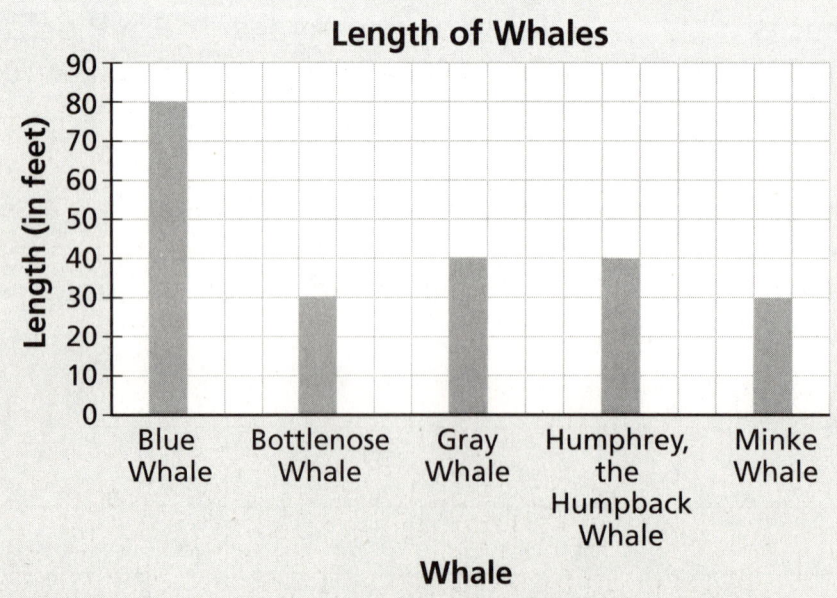

Length of Whales

Vocabulary — bar graph

26

▶ **Understand**   Plan   Try   Look Back

Use the newspaper article and the bar graph on page 26 to complete the sentences in exercises 24–30. ◀ MTK 273

24. Humphrey, the Humpback Whale, entered San Francisco Bay on October _____, 1985.

Did you know?
Humpback whales sing songs to each other. These songs can be heard underwater for 20 miles.

25. Humphrey swam _____ miles up the Sacramento River.

26. Humphrey is the same length as a typical _____ whale.

27. The Blue whale is _____ feet long.

28. The _____ and the _____ are each 30 feet long.

29. The longest whale in the graph is the _____.

30. The title of the bar graph is _____.

31. Use the information in the bar graph to write a sentence to compare whale lengths.

_____

_____

more ▶

27

Look at all the data and choose which facts to use.

It is so interesting to learn about *T. rex Sue* and Humphrey, the Humpback Whale. You decide to make a table to compare the facts and figures you've found.

T. rex Sue and Humphrey, the Humpback Whale

Characteristic	Sue	Humphrey
Average Length	42 feet	40 feet
Average Weight	7 tons	30 tons
Number of Teeth	58	0

Wow! You never realized that a whale like Humphrey has no teeth. You wonder if this is true for all whales. In a book about whales, you find the following graph.

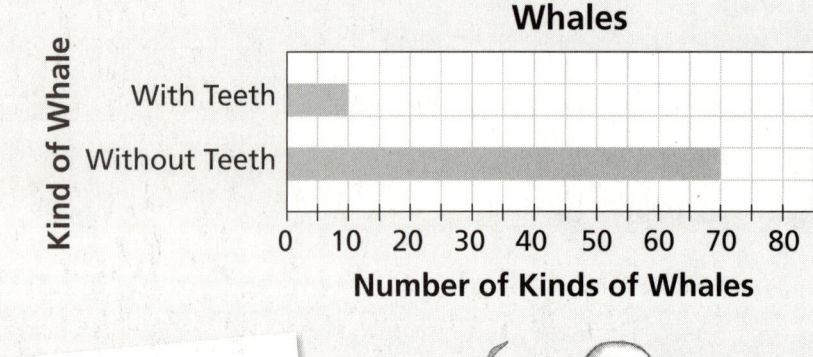

Did you know? Whales with no teeth are called *baleen whales*. These whales have plates that are called baleen inside their mouths. The tiny creatures that the whales eat are trapped inside the baleen. These creatures are so tiny that the whales don't need teeth for chewing.

▶ **Understand**   Plan   Try   Look Back

DO NOT solve exercises 32–34. Instead, write the information you would use to solve the problem. Then circle where you found the information. ◀ MTK 268, 273

32. How much more than *T. rex Sue* does Humphrey, the Humpback Whale, weigh?

To solve this problem, you need to find

a. the weight of *T. rex Sue*: _____7 tons_____

b. the weight of Humphrey, the Humpback Whale: _____30 tons_____

You got the information from the   (table)   bar graph

33. Who has more teeth, *T. rex Sue* or Humphrey, the Humpback Whale?

To solve this problem, you need to find

a. the number of teeth of *T. rex Sue*: _____

b. the number of teeth of Humphrey, the Humpback Whale: _____

You got the information from the   table   bar graph

34. How many more kinds of whales with teeth are there than without teeth?

To solve this problem, you need to find

a. the number of whales with teeth: _____

b. the number of whales without teeth: _____

You got the information from the   table   bar graph

29

Problems with Missing Information

Sometimes a word problem doesn't give you all the information you need.

To solve those word problems, you may be able to find the missing information yourself. For example, you might need to measure to find a height.

There are also word problems you won't be able to solve until you look up the missing information or until someone gives it to you.

Koko signs the word *hungry*.

Read each word problem. Then answer parts *a* and *b*.

Did you know?
Koko is a famous gorilla that speaks American Sign Language. This is the same language that is used by many people who cannot hear. Koko knows over 1000 signs.

1. Koko weighs 280 pounds. How much weight has Koko gained since she was born? ◂ MTK 358

   a. What information is missing from the word problem?

   _____

   b. Can you find the missing information? If *yes*, write the information and tell how you found it.

   _____

2. Koko stands 5 feet $\frac{3}{4}$ inch tall. What is the difference between Koko's height and your height? ◂ MTK 210–214, 346

   a. What information is missing from the problem?

   _____

   b. Can you find the missing information? If *yes*, write the information and tell how you found it.

   _____

▶ **Understand**   Plan   Try   Look Back

3. Today the elephants are going to work with only red and blue paint. Each elephant gets 2 cans of red paint and 1 can of blue paint. How many cans of paint are needed in all?

   a. What information is missing from the problem?

   _____

   b. Can you find the missing information? If *yes*, write the information and tell how you found it.

   _____

4. Humphrey, the Humpback Whale, was about 5 yards long when he was a baby. How many feet is that?   ◀ MTK 346

   a. What information is missing from the problem?

   _____

   b. Can you find the missing information? If *yes*, write the information and tell how you found it.

   _____

5. A Humpback whale like Humphrey blows a ten-foot-high fountain when it breathes. Is this higher or lower than your classroom ceiling?   ◀ MTK 336

   a. What information is missing from the problem?

   _____

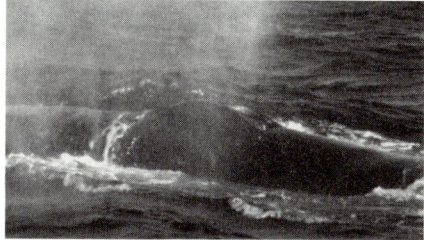

The spray you see is a Humpback whale breathing through its blowhole. A Humpback whale can hold its breath for up to 28 minutes.

   b. Can you find the missing information? If *yes*, write the information and tell how you found it.

   _____

more ▶

Vocabulary ▬ yard (yd)

31

There are many places to look for information.

Read each word problem. Then answer parts *a* and *b*.

6. The contestants for the next *The Animal Quiz Show* will be announced today at 4 P.M. How much longer do you have to wait to find out if you've been chosen? ◂MTK 339

   a. What information is missing from the problem?

   _____

   b. Can you find the missing information? If *yes*, write the information and tell how you found it.

   _____

7. Congratulations! You have been chosen to be on *The Animal Quiz Show*. There are 4 people who will compete for the grand prize. What fraction of these people are girls? ◂MTK 210–215

   a. What information is missing from the problem?

   _____

   b. Can you find the missing information? If *yes*, write the information and tell how you found it.

   _____

8. You have 20 seconds to answer a question on the show. You are in the middle of answering a question, how much time do you have left?

   a. What information is missing from the problem?

   _____

   b. Can you find the missing information? If *yes*, write the information and tell how you found it.

   _____

Vocabulary ▾ P.M. ▾ second (sec)

▶ **Understand**   Plan   Try   Look Back

On *The Animal Quiz Show*, you are asked the questions on this page.

Write the answer to each question. Then write the page number where you found that information.

9. Who are Arum, Desi, and Eva?

   _____ (page _____)

10. What is the name of one of the elephant paintings?

    _____ (page _____)

11. What kind of whale was Humphrey?

    _____ (page _____)

12. How long can a whale like Humphrey hold its breath?

    _____ (page _____)

13. Who is *T. rex Sue* named after?

    _____ (page _____)

14. How many signs does Koko the Gorilla know?

    _____ (page _____)

You are the First Prize Winner! Look back at page 18. Which prize will you choose?

_____

33

Chapter 2 Test

Fill in the circle with the letter of the correct answer.

1. Look at the table.

   According to the table, when did Lydia spend more time on homework than playing baseball?

   Ⓐ twice this week

   Ⓑ on Monday and Thursday

   Ⓒ Friday only

   Ⓓ none of the above

   Lydia's Week

Day of the week	Hours Doing Homework	Hours Playing Baseball
Monday	2	1
Tuesday	1	2
Wednesday	2	0
Thursday	1	1
Friday	0	3

2. Look at the bar graph.

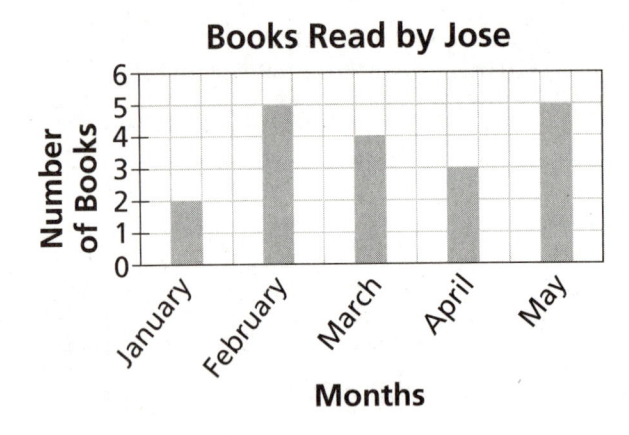

   Books Read by Jose

   (Bar graph showing: January = 2, February = 5, March = 4, April = 3, May = 5)

   In which months did Jose read more than 3 books?

   Ⓐ March only

   Ⓑ January and April

   Ⓒ April only

   Ⓓ February, March, and May

34

▶ **Understand**   Plan   Try   Look Back

Choose the letter of the best answer. Then write why you made that choice.

3. Roberta gives Janet 7 color markers. How many markers does Roberta have left?

   What information is needed to answer the problem?

   (A) the number of color markers Janet has in her collection

   _____

   (B) the date Roberta gave Janet the color markers

   _____

   (C) the number of color markers in a package

   _____

   (D) the number of color markers Roberta started with

   _____

Look at the pictograph.

4. Write 3 things this pictograph tells you about the number of tickets Lu sold.

   _____
   _____
   _____
   _____

Tickets Sold

Day of the Week	Number
Monday	🎟🎟🎟🎟🎟🎟🎟🎟
Tuesday	🎟🎟🎟🎟🎟🎟
Wednesday	🎟🎟🎟
Thursday	🎟🎟🎟
Friday	🎟

Key: 🎟 = 1 ticket

35

Chapter 3

Creative Inventions
Making a Plan

Inventors' Fair Coming Soon

The Inventors' Fair is next month. The Fair will have exhibits showing inventions from the past.

We invite you to put on your thinking caps and send in your idea for an invention at the Fair!

It is easy to enter your idea. Write a short description of your invention and draw a picture of it.

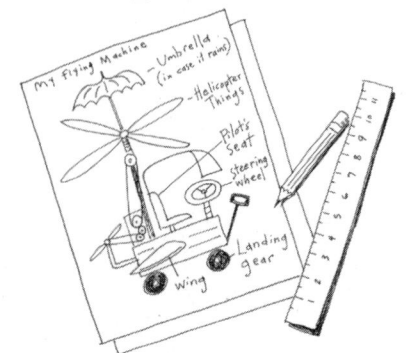

Understand ▸**Plan**   Try   Look Back

When inventors create something, they make mental pictures and write plans to help them get from start to finish. Making mental pictures and writing plans also help you solve math word problems. As you learn about inventions, you will practice using the first and second steps in the four-step problem-solving method: **Understand** and **Plan**.

▼ The first woman to get a patent in the United States was Hannah Slater. She got a patent in 1793 for cotton sewing thread.

▸ The first U. S. patent was given in 1790 to Samuel Hopkins who invented a formula to help make soap.

▲ Inventors apply for patents for their inventions. A patent says that the inventor owns the invention.

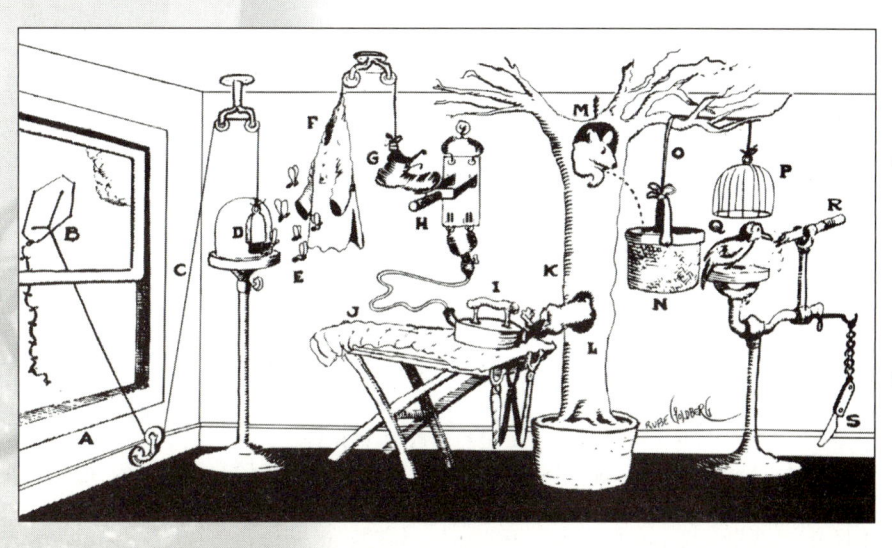

◂ Rube Goldberg was a famous cartoonist who drew his inventions. His cartoons always had lots of steps for simple everyday jobs, like sharpening your pencil.

37

Visualizing a Math Problem

Sometimes it helps to make a picture of a word problem.

You want to send in an idea for the Invention Fair. To get started, you read up on past inventions. You learn that the first moving robot was invented in 1936, that's years before you were born!

Moving robots need to be programmed to move from one spot to another. A picture can help you write a program or plan a path for the robot. How would the robot in the picture get to the refrigerator?

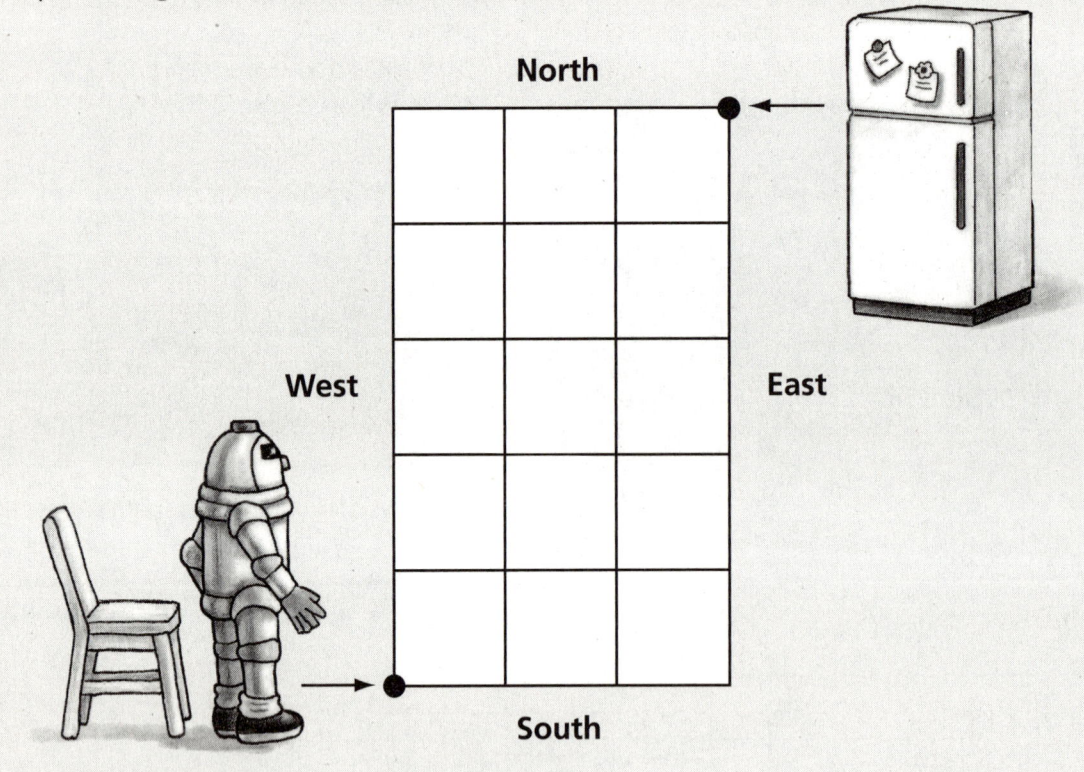

Here is one possible plan for the robot.

   Move 5 squares north.

   Move 3 squares east.

Understand ▸ Plan   Try   Look Back

To see some robots in action, you decide to visit the lab of an inventor who works with moving robots.

The inventor wants to see whether the robot can move a distance of 9 feet. The robot has moved 4 feet. How much further does the robot need to move?

When you read a math word problem, it helps to make a mental picture of the problem. You can imagine a robot standing along a path with a START and a FINISH marker at each end.

Or, you can use a number line to draw a simpler picture. You can count the spaces to see how far the robot has traveled and how far it has yet to go.

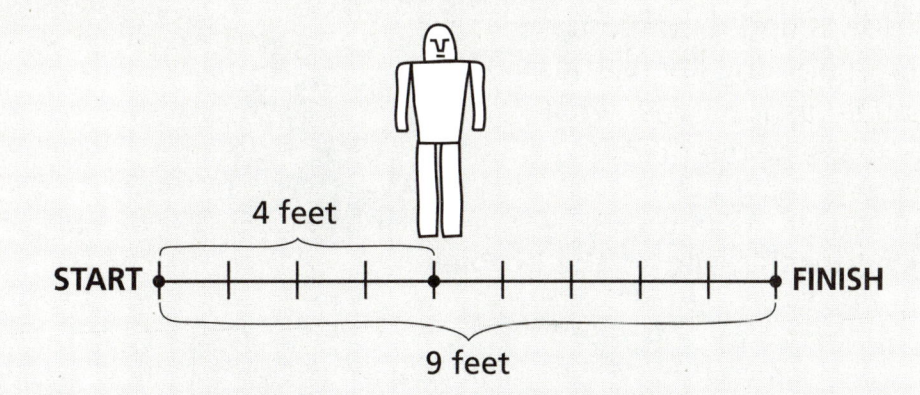

The picture helps you see that, in problems like this, you can subtract to find the answer. ◂ MTK 46–48

$$9 - 4 = 5$$

So, the robot needs to move 5 feet more.

more ▸

Vocabulary ▪ number line ▪ subtract

39

A good math picture can help you decide how to solve a word problem.

Read each word problem. Then follow the directions.

1. The inventor wants to know if the robot can move along the sides of a rectangular room that is 8 feet long and 6 feet wide. What is the distance around the room? ◀ MTK 36, 346, 349

   Circle the picture that shows the problem.

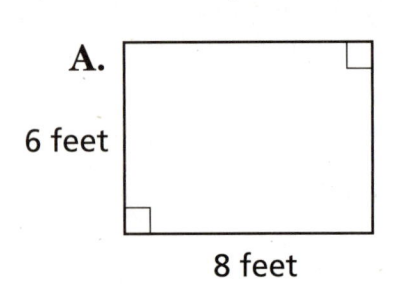

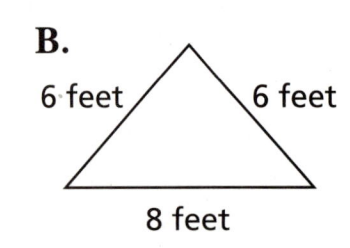

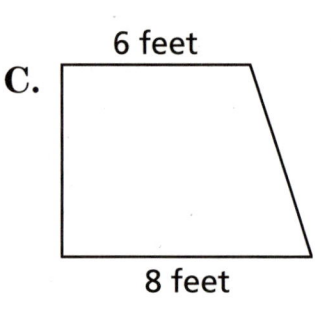

   A. 6 feet / 8 feet (rectangle)
   B. 6 feet / 6 feet / 8 feet (triangle)
   C. 6 feet / 8 feet (trapezoid)

   Circle one way to solve the problem.

   A. 8 + 6 + 8 + 6
   B. 8 + 6
   C. 8 + 6 + 6

2. The robot dog is tied to a 4-foot chain. If the robot dog keeps the chain stretched as it moves, what will be the shape of the dog's path? ◀ MTK 316, 346

   Circle the picture that shows the problem.

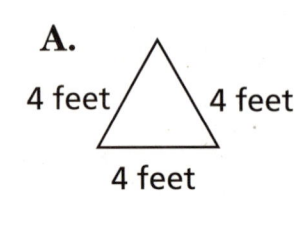

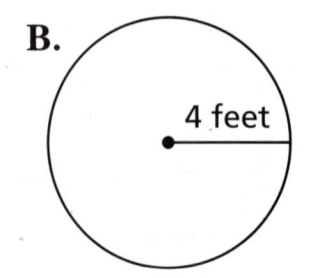

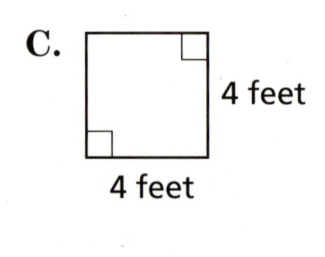

   A. 4 feet / 4 feet / 4 feet (triangle)
   B. 4 feet (circle)
   C. 4 feet / 4 feet (square)

   Circle the word that describes the shape of the dog's path.

   A. triangle
   B. rectangle
   C. circle

40 Vocabulary ▼ rectangle ▼ triangle ▼ circle

> Understand > Plan   Try   Look Back

You learn that robots are used for space exploration.

Did you know?
In 1997, a 12-year old girl won the contest to name the robot that explored Mars. She named the robot *Sojourner Truth* after the African-American woman who fought for women's rights and against slavery. *Sojourner* means traveler.

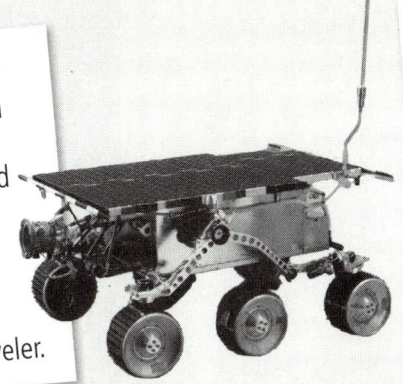

Did you know?
Scientists wanted to know more about the different rocks on Mars. *Sojourner* had a special instrument that carefully studied each rock. The scientists gave the rocks nicknames like *Barnacle Bill*, *Yogi*, and *Scooby Doo*.

3. *Sojourner* Rover found 14 different rocks. *Sojourner* explored each of these 14 rocks in detail. How many different rocks did *Sojourner* explore all together? ◂MTK 34–36

   Circle the picture that shows the problem.

   A. ○○○○○○○
      ○○○○○○○
      ○○○○○○○
      ○○○○○○○

   B. ⊗⊗⊗⊗⊗⊗⊗
      ○○○○○○○

   C. ○○○○○○○
      ○○○○○○○

4. The United States post office issued a postage stamp to honor *Sojourner* Rover. The postage stamp cost $3 each. How much did 4 of these stamps cost? ◂MTK 17, 60–62

   Circle the picture that shows the problem.

   A. $3 $3 $3 $3
      $3 $3 $3

   B. $3 $3 $3 $3

   C. ⊗ ⊗ ⊗ $3

   Circle one way to solve the problem.

   A. $4 + 3$ B. $4 \div 3$ C. 4×3

   more ▸

A good picture can lead to a good plan.

Did you know?
Rush Robinett got the idea for hopping robots from watching grasshoppers.

Read each word problem. Then follow the directions.

5. At Sandia Labs, home of the famous hopping robots, you watch a robot-hopping contest. Robot *X* hops 30 feet high. Robot *Y* hops 21 feet high. How much higher did Robot *X* hop? ◀ MTK 161, 346

   Draw a picture that shows the problem.

   Circle one way to solve the problem.

   A. 30 + 21

   B. 30 − 21

   C. 30 × 21

   D. 30 ÷ 21

6. In another demonstration, you watch Robot *W* travel 6 feet with each hop. How far can it travel in 3 hops? ◀ MTK 64, 346

   Draw a picture that shows the problem.

   Circle one way to solve the problem.

   A. 6 + 3

   B. 6 − 3

   C. 6 × 3

   D. 6 ÷ 3

▶ **Understand** ▶ **Plan** Try Look Back

Use a simpler word problem to plan your solution.

You can change the numbers in a word problem to make it easier to plan a solution. Then, go back and solve the original problem.

Example
Original problem A robot moved 29 feet to a chair. Then it moved another 32 feet to a door. How far did it move?

Think!

Write a simpler problem A robot moved 2 feet to a chair. Then it moved 3 feet to a door. How far did it move from start to finish?

Picture the simpler problem

2 feet   3 feet

How to solve the simpler problem
total distance moved = distance to chair + distance to door

How to solve the original problem Add: 29 + 32

Think of a simpler problem first.

7. A robot moved along the sides of a square room. Each side was 20 feet. How far did the robot move in all? ◀ MTK 62, 346, 349

Think!

Write a simpler problem A robot moved around a square room that was 2 feet on each side. How far did the robot move in all?

Picture the simpler problem

2 feet
2 feet   2 feet
2 feet

Circle a number sentence that shows how to solve the simpler problem. ◀ MTK 36, 255

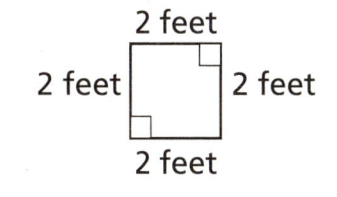

A. total distance = length of side × 4
B. total distance = length of side + length of side
C. total distance = length of side + 4

Circle a number sentence that shows how to solve the original problem.

A. total distance = 20 × 4
B. total distance = 20 + 20
C. total distance = 20 + 4

more ▶

Vocabulary ▼ add ▼ sum ▼ total ▼ number sentence

Your simpler word problem can be different from a friend's.

Did you know?
One of the first robots invented was named *Elektro* and its dog's name was *Sparko*. People got to see them at the World's Fair in 1939.

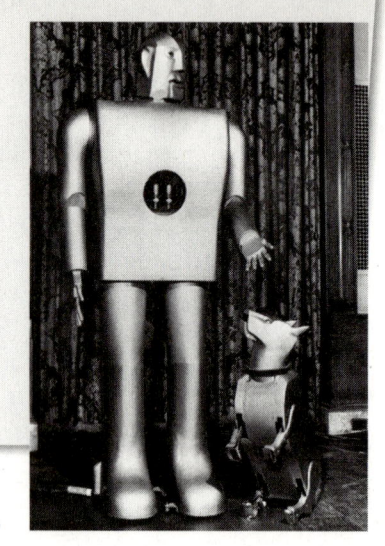

Read each word problem. Then follow the directions.

8. *Elektro* could say 77 different words. You are making a list. How many more are left after it has said 43 different words? ◀ MTK 160

Think!

Write a simpler problem *Elektro* knows 5 different words. How many more words are left after it said 3 different words?

Picture the simpler problem.

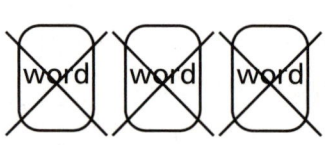

Circle a number sentence that shows how to solve the simpler problem.

A. number of words left = total number of words + number of words said

B. number of words left = total number of words − number of words said

C. number of words left = number of words said + total number of words

Circle a number sentence that shows how to solve the original problem.

A. number of words left = 77 + 43

B. number of words left = 77 − 43

C. number of words left = 43 + 77

44

▶ **Understand** ▶ **Plan** Try Look Back

Robots are really interesting. Now you want to learn about another invention.

Did you know? Crayons were invented in 1903.

9. According to experts, an average child will use about 730 crayons by age 10. About how many crayons will 2 children use by the time they are 10 years old? ◀ MTK 146–149

Think!

Write a simpler problem.

_____

_____

_____

Draw a picture of the simpler problem.

Write a number sentence that shows how to solve the simpler problem.

_____

There are different ways to solve a word problem.

Circle all the number sentences that show how to solve the original problem.

A. number of crayons = 730 + 730

B. number of crayons = 730 + 2

C. number of crayons = 730 × 2

D. number of crayons = 730 ÷ 730

Vocabulary ▬ average

45

Making a Plan

Plans can have two or more steps.

When inventors make a plan, the plan often has more than one step.

Plans for solving a word problem can also have more than one step.

Circle the plan that you could use to solve the word problem.

1. The first box of crayons sold for a nickel. How many 5-cent boxes of crayons could you buy with 1 quarter and 2 dimes? ◀ MTK 74–76, 17

   Plan A
   - Find how much money 1 quarter plus 2 dimes equals.
   - Multiply that sum by 5.

   Plan B
   - Find how much money 1 quarter plus 2 dimes equals.
   - Divide that sum by 5.

Vocabulary ▪ nickel ▪ quarter ▪ dime ▪ plus ▪ multiply ▪ divide

▶ **Understand** ▶ **Plan** Try Look Back

2. The first box of crayons had 8 different colors. Later, 2 new colors were added. Today you can buy a box with 120 different colors. How many more colors have been created since the two new colors were added?

   Plan A
   - Add 8 and 2.
   - Subtract the sum from 120. ◀ MTK 161

   Plan B
   - Subtract 8 from 120.
   - Add 2 to the difference.

   Did you know?
   The colors in the first box of crayons were black, brown, blue, red, purple, orange, yellow, and green.

3. There is a box of crayons pictured on a 32¢ postage stamp. How many of these stamps could you buy with $1? ◀ MTK 19, 161–163

   Plan A
   - Change $1 to 100¢.
   - Subtract 32¢ from 100¢.
   - Keep subtracting 32¢ until you have a difference that is less than 32¢.
   - Count the number of times you subtracted 32¢.

   Plan B
   - Change $1 to 100¢.
   - Add 32¢ and 100¢.

   Plan C
   - Change $1 to 100¢.
   - Multiply 32¢ and 100¢.

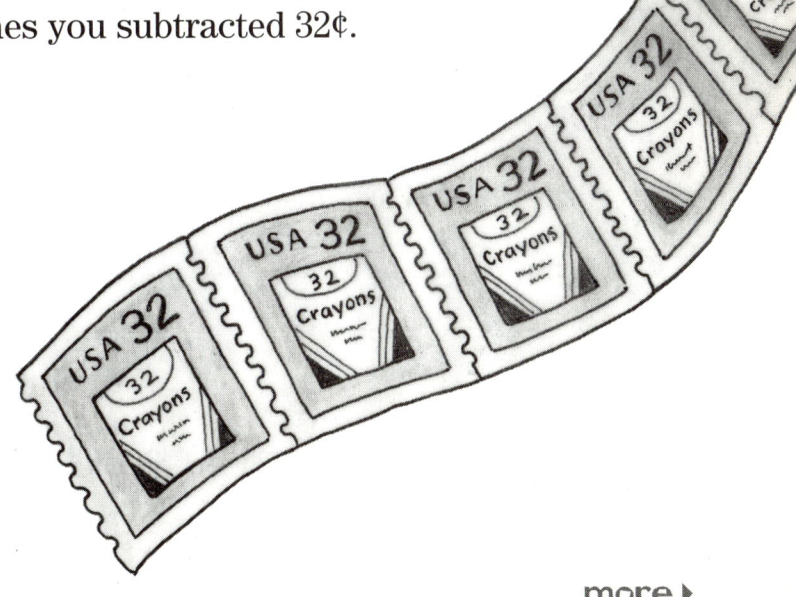

more ▶

Vocabulary ▪ first ▪ difference ▪ cent (¢)

47

A plan can work if it uses the right information.

You just heard on the radio about a competition called *RoboCup*. Curious to find out more, you logon to the competition website. You learn that it is an international soccer tournament for robots. *Incredible!*

You find a bar graph listing the number of countries that have participated in the *RoboCup* competitions.

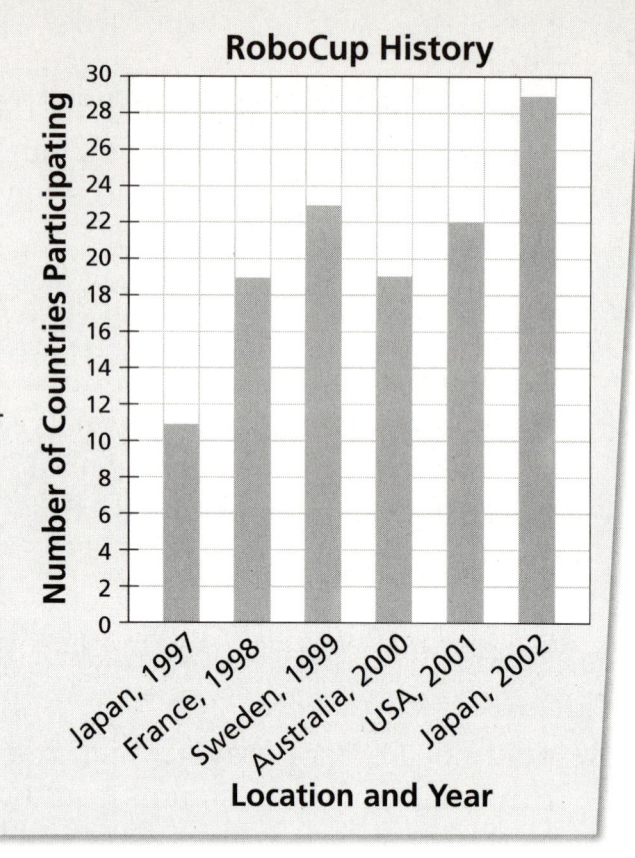

RoboCup History

(Bar graph: Number of Countries Participating vs Location and Year)
- Japan, 1997: 11
- France, 1998: 19
- Sweden, 1999: 23
- Australia, 2000: 19
- USA, 2001: 22
- Japan, 2002: 29

For exercises 4–5, use the information from the bar graph. Circle the plan that you could use to solve the word problem.

4. How many more countries participated in 2002 than in 1997? ◀ MTK 160, 273

   Plan A
   - Find the number of participants in each of the two years.
   - Subtract 1997 from 2002.

   Plan B
   - Find the number of participants in each of the two years.
   - Subtract 11 from 29.

5. Which year had twice the number of participants as in 1997? ◀ MTK 172–173, 273

   Plan A
   - Multiply 11 by 2.
   - Find the year with that many participants.

   Plan B
   - Divide 11 by 2.
   - Find the year with that many participants.

Did you know?
The goal of *RoboCup* competition is to design fully independent humanoid robots by the year 2050. Designers hope that these robots will be able to play against human world soccer champion teams.

Vocabulary ▼ twice

▶ Understand ▶ Plan Try Look Back

Completing a plan can help you write one of your own.

Complete the plan. Fill in each blank with a word from the box.

6. Two robots on a soccer field are painted with stripes and the other robot is painted with dots. What fraction of the robots are painted with stripes? ◀ MTK 34–36, 210–212

> add
> subtract
> numerator
> denominator

Plan

- _____Add_____ 2 and 1.

- Write the sum in the _____ of the fraction.

- Write 2 in the _____ of the fraction.

7. The robots play two 10-minute halves to make a game. It is now 16 minutes into the game. How much time is left in the game? ◀ MTK 146–147, 334

> add
> subtract
> sum
> difference

Plan

- _____ 10 and 10.

- _____ 16 from the _____.

8. If a robot in a soccer game holds the ball for too long, that robot has to leave the field for a 30-second penalty. The Green Team robot has been off the field for 23 seconds. In how many seconds can the robot return to the field? ◀ MTK 161, 337

> add
> subtract
> multiply
> divide

Plan

- _____ 23 from 30.

Did you know?
RoboCup Junior is a special category started in 2000. It is designed just for middle school and high school students.

more ▶

Vocabulary ▪ half ▪ numerator ▪ denominator

49

Making a good plan is a key to solving word problems.

Write a plan for solving each word problem.

9. Thomas Edison invented the light bulb. In 1879, he got a light bulb to glow for 40 hours. Could that bulb last from Monday at 8 A.M. to Wednesday at 8 A.M., 2 days later? ◀ MTK 146–147, 12–13, 334–341

Did you know?
Thomas Edison had more than 1000 patents for his inventions.

Plan

_____

_____

10. By the time Beulah Henry was 26 years old, she had 3 patents. She received 46 more patents during her lifetime. How many patents did she receive in all? ◀ MTK 146–147

Did you know?
Beulah Henry's first invention was an ice cream freezer. She invented so many things that she was often called *The Lady Edison*.

Plan

_____

_____

11. Before 1883, a shoemaker could make about 50 shoes in a day. With Jan E. Matzeliger's machine, that number jumped to about 700 shoes in a day. About how many more shoes could a shoemaker make in a day with Matzeliger's invention? ◀ MTK 161–163

Jan E. Matzeliger invented a machine to make shoes in 1883.

Plan

_____

_____

50   Vocabulary ▼ hour (h) ▼ A.M.

Understand ▸ Plan Try Look Back

12. A firefighter could breathe fresh air from the bag on Morgan's helmet for about 15 minutes. If a firefighter had been breathing from the helmet from 11:05 A.M. to 11:15 A.M., about how many more minutes of air were left in the helmet's bag? ◂ MTK 46–58, 338–339

Did you know?
Garrett A. Morgan invented a traffic signal and a fire safety helmet. Both of these inventions are displayed at Detroit's Museum of African American History.

Plan

_____

_____

13. A pair of Chester Greenwood's earmuffs cost 25¢. A friend bought 1 pair of earmuffs and his sister bought 2 pairs. Together they paid Chester $1. How much change did they get back? ◂ MTK 38–39, 46–58

Plan

_____

_____

When Chester Greenwood's ears got cold, he needed something to keep them warm. So, he invented earmuffs. What could you use in your life that doesn't exist yet? Now it's your chance to do some inventing!

Remember, once you get your invention idea, a good plan will help you make your idea come true. If you make and follow a good plan, you will succeed in solving math word problems, too.

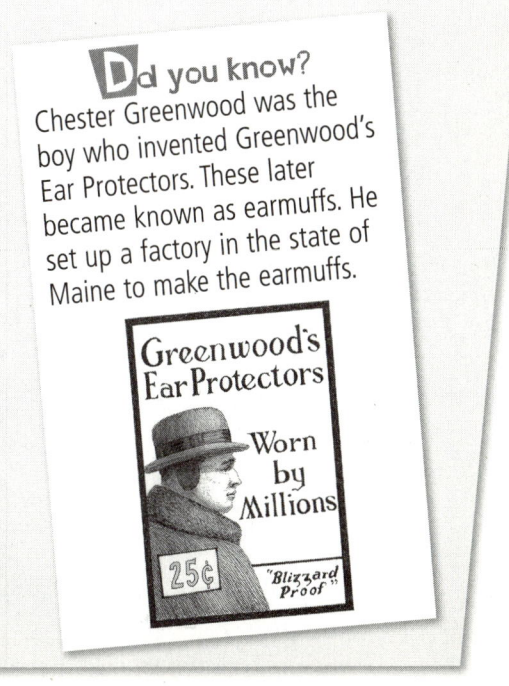

Did you know?
Chester Greenwood was the boy who invented Greenwood's Ear Protectors. These later became known as earmuffs. He set up a factory in the state of Maine to make the earmuffs.

Greenwood's Ear Protectors
Worn by Millions
25¢ "Blizzard Proof"

51

Chapter 3 Test

Fill in the circle with the letter of the correct answer.

1. The florist had 12 miniature rose plants. Joan bought 5 of them. Which picture best shows the number of plants left?

    Ⓐ ||||||||||| ||||||

    Ⓑ ||||||| ✗✗✗✗✗

    Ⓒ [||||| |||||] [||||| |||||] [||||| |||||] [||||| |||||] [||||| |||||]

2. There are 3 people at the airport. Each person has 2 pieces of luggage. Which picture best shows the total pieces of luggage they have?

    Ⓐ || || Ⓑ ||| || Ⓒ || || ||

3. The circus performer walked 3 feet across the tightrope and stopped for 10 seconds. Then she walked another 5 feet. Which picture best shows how far she walked in all?

    Ⓐ 5 feet / 3 feet (overlapping)

    Ⓑ 3 feet, 5 feet (separate)

    Ⓒ 3 feet, 5 feet (separate)

4. A party of 15 people enter a restaurant. They are seated equally at 3 different tables. Which pictures shows how they are seated?

    Ⓐ (tables of 3, 3, 2)

    Ⓑ (tables of 5, 5, 5)

    Ⓒ (tables of 3, 3, 3, 3, 3)

52

Fill in the circle with the letter of the correct answer. Write the reason why you made your choice.

5. There are some CDs on a rack. Three of the CDs are for data and 2 are for music. How could you compute the total number of CDs on the rack?

    Ⓐ 3 + 2 _____

    Ⓑ 3 − 2 _____

    Ⓒ 3 × 2 _____

    Ⓓ 3 ÷ 2 _____

For questions 6–8, write your answer on the lines provided.

6. Roberto buys 3 pens. Each pen costs $2. Write the steps you would use to find the total cost of the pens.

    _____

    _____

7. Juanita has $10. She spent $4.75 on a movie ticket and $2.50 on popcorn. Write the steps you would use to find how much money she has left.

    _____

    _____

8. Read the plan. Then write a math word problem that could be solved by using the plan.

    Plan
    - Add 8 and 5.
    - Subtract the sum from 18.

    _____

    _____

Chapter 4

Contests and Races

Carrying Out the Plan

Books by Kids for Kids

Dear _____,
(Your Name)

I have heard that you like contests and races. How would you like to help me write a book for kids about contests and races? I will send you all over the world to learn about interesting events. You will even get to enter some of them so that you can help me write the book. Please get back to me as soon as you can. I want to start this project right away.

Sincerely,

Sam Lee

Sam Lee
Editor-in-Chief

Understand ▶Plan ▶Try Look Back

As you learn all about contests and races, creating and using a plan can make you a winner. In this chapter, you'll be making plans to solve word problems. You'll practice using the second and third steps of the four-step problem solving method: **Plan** and **Try**.

▸ When the Dutch first came to the New World, they played a game using two jump ropes. Since the Dutch brought it to America, people began calling it *Double Dutch*. Today there are *Double Dutch* contests all over the United States.

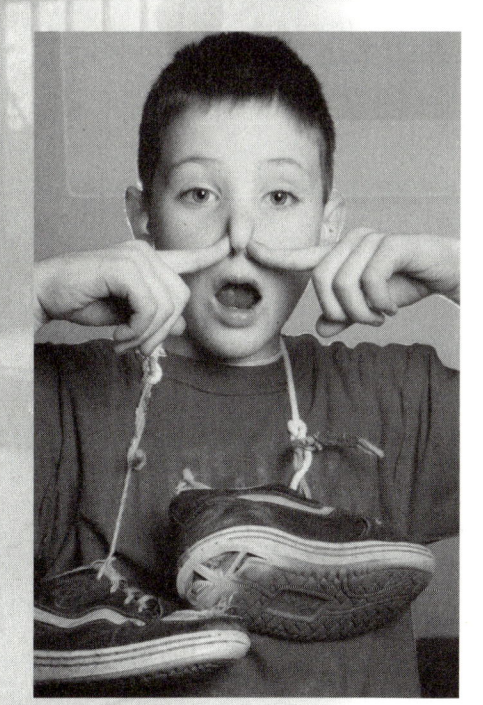

▴ In 1976, a storeowner in Vermont took a picture of a kid with a very rotten pair of sneakers to use in an advertisement. That started *The Rotten Sneaker Contest*. This is one contest where the lowest score wins! Above is the picture of a past winner.

◂ There is a *National Children's Whistling Championship* in North Carolina every year. Contestants must whistle three different tunes.

55

Choosing the Correct Solution

Sometimes drawing a diagram of the problem can help you come up with a plan.

Your first assignment for the book is to learn about the *Kinetic Sculpture Race*.

Did you know?
In 1969, an artist in Ferndale, California, decided to make his son's tricycle into a sculpture that moved. He liked it so much he decided to host a race each year. The race became the Kinetic Sculpture Race.

Solve the word problems by completing each step. The first exercise is done for you.

1. You enter *The Kinetic Sculpture Race*. Your sculpture has 5 poles. You want to use 20 stars to decorate the poles. How many stars will be on each pole, if you put the same number of stars on each pole? ◀ MTK 74–76

Plan Use a picture to help you make a plan. Circle the diagram that is more helpful.

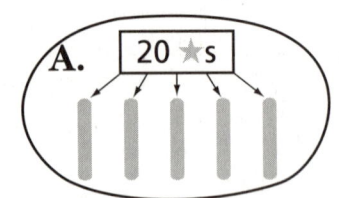

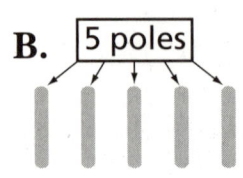

A. (circled) 20 ★s → 5 poles

B. 5 poles → poles

Fill in the blanks in the plan.

- Divide the number of __stars__ by the number of __poles__.

Try Circle the work that fits the plan.

A. 20 ÷ 5 = 25   B. (circled) 20 ÷ 5 = 4   C. 20 − 5 = 15

There will be __4__ stars on each pole.

56   **Vocabulary** ◀ solution

Understand ▶Plan ▶Try Look Back

2. Teams are penalized if rules are broken. You get a penalty for not guarding your sculpture at night. You get a second penalty for asking a passing boat for help. Finally, you get a penalty for asking help from a passing truck. Each penalty is worth 3 hours. How many hours of penalties do you have?

Plan Use a picture to help you make a plan. Circle the diagram that is more helpful.

A. [3 hours] ←→ [3 hours]   B. [3 hours] + [3 hours] + [3 hours]

Fill in the blanks in the plan.

- Skip count by _____ to find the total hours of penalty.

Try Circle the work that fits the plan.

A. 1, 2, 3   B. 3, 6, 9   C. 3, 4, 5

Complete the answer. You have _____ hours of penalties.

---

3. The 3-day race is 38 miles long. You went 15 miles on day 1. You have gone 7 miles so far on day 2. How much further before you finish? ◀ MTK 140, 160, 346

> **Did you know?**
> The *Kinetic Sculpture Race* is held over Memorial Weekend. The sculptures race over road, water, sand, and mud!

Plan Use a picture to help you make a plan. Circle the diagram that is more helpful.

A. |—15—|—7—|—?—| B. |—15—|—7—|
   0   38   Start   Finish

Fill in the blanks in the plan.

- Add _____ and _____ to find the distance covered.

- Subtract the sum from _____.

Try Circle the work that fits the plan.

A. 15 + 7 = 22   B. 38 − 15 = 23   C. 15 + 7 = 22
   38 − 22 = 16   23 − 15 = 8   38 + 22 = 60

Complete the answer. You have _____ miles more to go.   more ▶

Vocabulary ▼ skip count

When carrying out your plan, make sure that you use the correct information.

A friend just told you about another popular event. This is the *Hot Dog Eating Contest* held each year on July 4 on Coney Island in New York.

Solve the word problems by completing each step.

4. In 2002, the winner of the *Hot Dog Eating Contest* ate about 51 hot dogs. The next year, he won again by eating about 44 hot dogs. How many fewer hot dogs did he eat in 2003? ◀ MTK 161

Plan Fill in the blanks in the plan.

- Subtract _____ from _____.

Try Circle the work that fits the plan.

    A.   51   **B.**   2003   **C.**   2003
        −44   − 51   −2002
          7   1952   1

Complete the answer. He ate _____ fewer hot dogs in 2003.

5. The *Hot Dog Eating Contest* had a 12-minute time limit. Suppose they shortened the contest to 4 minutes. How many times longer would the old contest be compared to the new one? ◀ MTK 74–75, 76

Plan Fill in the blank in the plan.

- Divide 12 by _____.

Try Circle the work that fits the plan.

    A. 12 ÷ 3 = 4   **B.** 12 ÷ 4 = 3   **C.** 12 × 4 = 48

Complete the answer. The old contest would be _____ times longer.

Understand ▶Plan ▶Try Look Back

Some friends heard about the book you are working on. They invite you to join the *Spaghetti Bridge Building Contest.*

6. Your spaghetti bridge is finally built. Now you wait to see if it can hold a 2-kilogram weight for 5 minutes. The clock shows that 3 minutes have passed. How much longer do you have to wait? ◀ MTK 46–48, 359

Did you know?
The *Spaghetti Bridge Building Contest* takes place in March. Elementary and high school students build bridges using pasta. Now, that's using your noodles!

Plan Fill in the blanks in the plan.

- _____ 3 from _____.

Try Circle the work that fits the plan.

A. 5 − 2 = 3   **B.** 5 − 3 = 2   **C.** 5 + 3 = 7

Complete the answer. You need to wait for _____ minutes more.

Another contest your friends mentioned was the *National Children's Whistling Championship.*

7. You decide to enter the whistling championship. You need to practice 1 hour daily. But, you can practice only 20 minutes at a time. How many times do you need to practice each day? ◀ MTK 60–62

Plan Fill in the blank in the plan.

1 hour = 60 minutes

Think!
- Skip count by _____ up to 60.
- _____ is the number of skip counts.

Try Circle the work that fits the plan.

A. 20, 40, 60   **B.** 2, 4, 6   **C.** 2, 2, 2

Complete the answer. You need to practice _____ times each day in order to reach 1 hour.

more ▶

Vocabulary ▪ kilogram (kg)

59

Plans can have one, two, or many steps.

You decide to go to Japan to enter in a kite-flying contest.

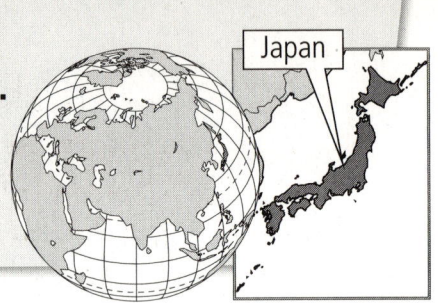

Solve the word problems by completing each step.

8. A kite-maker is making 4 kites for you. It takes 40 minutes to hand paint a kite. Is 2 hours enough time to paint 4 kites? ◀ MTK 12, 60–62

Plan Fill in the blanks in the plan.

Think! 2 hours = 120 minutes

- _____ 40 by _____.

- Compare the answer to 120.

Did you know?
Traditional Japanese kites are hand-painted using special pigments, mostly reds, blues, and purples.

Try Circle the work that fits the plan.

  A. $40 \times 4 = 160$
     $160 > 120$

  B. $40 - 4 = 36$
     $36 < 120$

  C. $40 \div 4 = 10$
     $10 < 120$

Complete the answer. It will take _____ than 2 hours.

---

9. At the kite-maker's shop, you see 3 purple kites and 7 red kites. What fraction of the kites are purple? ◀ MTK 212–213

Plan Fill in the blanks in the plan.

- Add 3 and _____.

- Write _____ in the denominator and _____ in the numerator.

Try Circle the work that fits the plan.

  A. $3 + 7 = 10$, $\frac{10}{3}$
  B. $3 + 7 = 10$, $\frac{3}{10}$
  C. $3 + 4 = 7$, $\frac{3}{7}$

Complete the answer. _____ of the kites are purple.

10. It takes a team of 35 people to fly one of the competition kites. There are 7 teams in this round. Are there more than 200 people involved? ◂MTK 172–175

Plan Fill in the blanks in the plan.

- _____ 35 by _____.

- Compare the answer to 200.

Try Circle the work that fits the plan.

A. 35 − 7 = 28
28 < 200

B. 35 × 7 = 245
245 > 200

C. 35 ÷ 7 = 5
5 < 200

Complete the answer. There are _____ than 200 people involved.

---

11. The contest sponsors will host a party tonight. They expect 300 guests. The sponsors want to serve sushi. Each sushi platter feeds 50 people. How many platters should they order? ◂MTK 82

Plan Fill in the blanks in the plan.

- Subtract 50 from _____.

- Keep subtracting _____ until the answer equals 0.

- Count the number of times you subtracted by _____.

Try Circle the work that fits the plan.

A. 300 − 50 = 250 → 250 − 50 = 200 → 200 − 50 = 150 →
150 − 50 = 100 → 100 − 50 = 50 → 50 − 50 = 0
I subtracted 50 six times.

B. 50 − 50 = 0   I subtracted 50 one time.

C. 300 − 50 = 250   I subtracted 50 one time.

Complete the answer. They need to order _____ platter(s) of sushi.

Using a Plan to Solve a Word Problem

After you make a plan it is important to try it out.

After attending the *Spider Beauty Contest* you decide to conduct a similar contest with your friends. The pictograph shows how your friends voted.

Did you know? A *Spider Beauty Contest* is held at the Kumo Gassen Festival each year in Japan.

Most Beautiful Spider

Name of Spider	Number of Votes
Daddy Long Legs Spider	✓ ✓ ✓ ✓ ✓
Trap Door Spider	✓ ✓ ✓ ✓
Jumping Spider	✓ ✓
Wolf Spider	✓ ✓ ✓

Key: ✓ = 1 vote

Look at the pictograph. Read each word problem. Then follow the directions.

1. Each friend voted once for the most beautiful spider. How many friends voted in all? ◂ MTK 36, 270–271

Plan Fill in the blanks in the plan.

- Count the number of _____ for each spider.

- _____ the votes to find the total.

Try Use pictures, numbers, or words to show your work.

Write your answer as a complete sentence.

_____

62

Understand ▶Plan ▶Try Look Back

2. Which spider got twice as many votes as the Jumping spider? ◂MTK 60–62, 270–271

Plan Fill in the blanks in the plan.

- The Jumping spider got _____ votes.
- Multiply _____ by 2.
- Find the spider with the same number of votes as the product.

Try Use pictures, numbers, or words to show your work.

Write your answer as a complete sentence.

_____

_____

3. One friend changes her vote. She votes for the Trap Door spider instead of Daddy Long Legs spider. How does that change the winner? ◂MTK 12, 270–271

Plan Fill in the blank in the plan.

- The new count for the Trap Door spider is _____ votes.
- The new count for the Daddy Long Legs spider is _____ votes.
- Compare the numbers to find the results.

Try Use pictures, numbers, or words to show your work.

Write your answer as a complete sentence.

_____

more ▶

Sometimes word problems have too little or too much information.

Did you know?
The *Great Crate Race* takes place in Maine every year. The racecourse is a string of lobster crates between two piers in the Atlantic Ocean. Contestants run back and forth over the crates until they fall in. The winner is the person who goes over the most crates.

The Great Crate Race Weight Divisions

Division	Weight of Runner
Lightweight	100 pounds and under
Middleweight	101–150 pounds
Heavyweight	151–200 pounds
Ultra-Heavyweight	201 pounds and above

Read each word problem. Then follow the directions.

4. Your friends Sam and Kevin are in the *Great Crate Race*. Sam weighs 63 pounds. Kevin weighs 50 pounds more than Sam. Which division will Kevin be in? ◂ MTK 268, 358

Plan Fill in the blanks in the plan.

- Add _____ and _____ to find Kevin's weight.

- _____ Kevin's weight with the weight divisions listed in the table.

Try Use pictures, numbers, or words to show your work.

Write your answer as a complete sentence.

64

Understand ▸ Plan ▸ Try   Look Back

You find another interesting competition. Each year the Canadian city of Nanaimo holds a *Bathtub Racing and Marine Festival*.

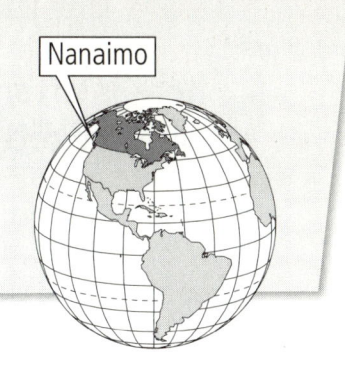

Nanaimo

5. You come in fifth in the bathtub race. The table shows the points for each place. The number of points for fifth place is smudged. How many points will you get for placing fifth? ◂ MTK 16, 46–48, 268

Bathtub Race

Place	Points
First Place	50 points
Second Place	48 points
Third Place	46 points
Fourth Place	44 points
Fifth Place	oints

Plan Fill in the blanks in the plan.

Think! Look for a pattern in the table.

- Find the _____ from one place to another.

- Count backwards by _____ from 50 four times.

Try Use pictures, numbers, or words to show your work.

Write your answer as a complete sentence.

_____

Did you know? The first bathtub race was held in 1967. Close to 200 tubbers entered the competition.

65

Writing the Plan and Solving the Word Problem

There is often more than one good plan for solving a word problem.

You just learned about *The Refrigerator Art Contest* that is on the Internet. Each week, five pieces of refrigerator artwork are displayed on the contest website. Then viewers vote for their favorite. This sounds interesting, so you decide to send in a picture.

Did you know? Votes for the best refrigerator picture come in from around the globe.

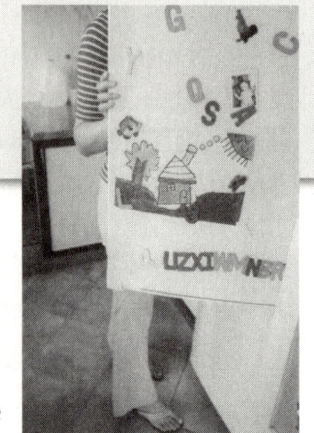

Read each word problem. Then follow the directions.

1. The producers call to tell you that your picture has been chosen. It is displayed on the website with 4 others. You log on to the website. Who are the top 3 vote-getters? ◀ MTK 14–15, 273

Plan Write a plan.

_____

_____

_____

Refrigerator Art Vote

Bar graph — Number of Votes by Contestant:
- (Your Name): 60
- Alex: 40
- Emma: 70
- Lillian: 20
- Sarah: 30

Try Show your work.

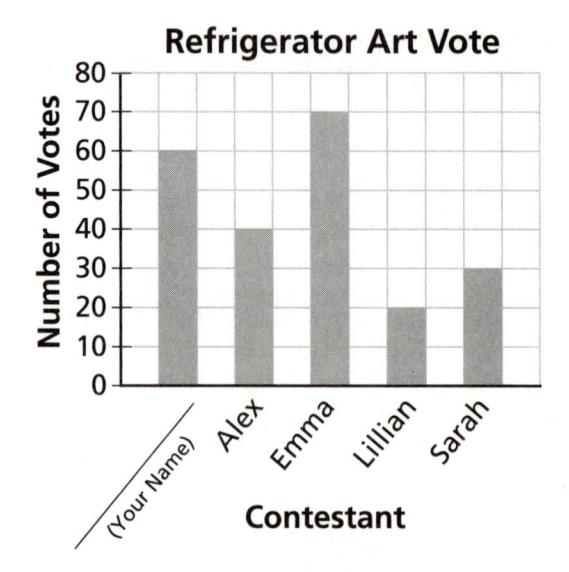

Write your answer as a complete sentence.

_____

66

Understand ▸**Plan** ▸**Try** Look Back

Your cousin tells you about the *Cannon Beach Sandcastle Contest.*

In 1964, a giant wave called a *tsunami* washed out the bridge to Cannon Beach, Oregon. People were stuck. Families couldn't go anywhere, so they decided to spend the day building sandcastles. That is how the *Cannon Beach Sandcastle Contest* started.

Did you know?
Only the first 150 entries are accepted to compete each year.

2. Spending a day at the beach sounds great. You invite friends to join you in the *Cannon Beach Sandcastle Contest.* So far, there are 29 of you. The rules allow 8 people on a team. How many more people do you need to make 4 teams? ◂ MTK 46–47, 62, 161, 172

Plan Write a plan.

_____

_____

_____

_____

Try Show your work.

Write your answer as a complete sentence.

_____

more ▸

If one plan doesn't work, think of another way to find the answer. Don't give up.

Read each word problem. Then follow the directions.

3. It costs $5 for each person to enter the *Cannon Beach Sandcastle Contest*. You have raised $30. How much more money do you need for a team of 8 people to enter? ◀ MTK 17, 60-62

Plan Write a plan.

_____

_____

_____

_____

Try Show your work.

Write your answer as a complete sentence.

_____

Understand ▶ **Plan** ▶ **Try** Look Back

4. Your team wins the sandcastle contest! The reward is a rectangular cake. It measures 8 inches long and 4 inches wide. You will cut the cake into 8 equal pieces. What will be the length and width of each piece? ◀ MTK 313, 346

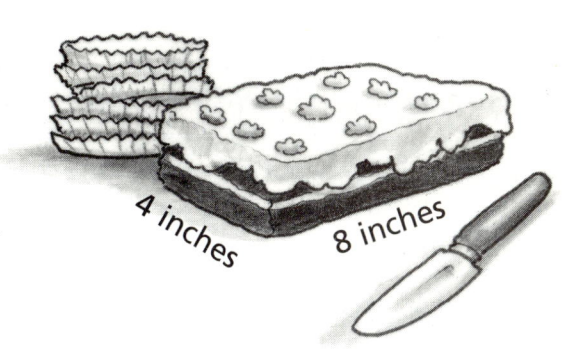

Plan Write a plan.

_____
_____
_____
_____

Try Show your work.

Write your answer as a complete sentence.

_____
_____
_____

You are now ready to help write the book about contests and races. You will make a suggestion to the Editor-in-Chief about using a picture of your sandcastle for the book cover.

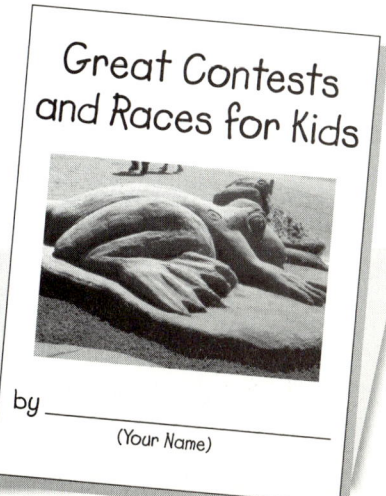

Great Contests and Races for Kids

by _____
(Your Name)

69

Chapter 4 Test

Fill in the circle with the letter of the correct answer.

1. Lucinda planted 3 rows of tomatoes. She has 6 tomato plants in each row. How can you find the number of tomato plants she planted in all?

   (A) Subtract 3 from 6.   (C) Add 3 and 6.

   (B) Multiply 3 by 6.   (D) Divide 6 by 3.

2. Juan bought 2 books. One cost $8 and the other $5. He gave the clerk a twenty-dollar bill. How can you find the amount of change he received?

   (A) Add 2, 8, and 5.
   Subtract the answer from 20.

   (C) Add 8 and 5.
   Subtract the answer from 20.

   (B) Add 8 and 5.
   Add the answer to 20.

   (D) Subtract 8 from 20.
   Add 5 to the answer.

3. Leah wants to build a doghouse. She needs a piece of board that is 3 feet long and another that is 9 feet long. Both pieces are 1 foot wide. If she buys one board to cut into the two pieces, how long should the board be?

   (A) between 3 feet and 9 feet

   (B) 12 feet or longer

   (C) shorter than 9 feet

   Write or draw your plan here.

Understand ▶Plan ▶Try Look Back

Choose the letter of the best answer. Then write why you made that choice.

4. Jamille is on an 18-mile hike with her friends. The first day they hiked 5 miles and the next day they went the same distance. How many more miles do they have to hike?

   (A) 23 miles _____

   (B) 13 miles _____

   (C) 8 miles   _____

5. Nathan has a collection of 3 toy cars and 5 toy planes. What fraction of his collection are the toy cars?

   (A) $\frac{3}{8}$ _____

   (B) $\frac{3}{5}$ _____

   (C) $\frac{5}{8}$ _____

Write your plan and show your work

6. Ken wants to put a fence around his garden. His garden is a rectangle that is 8 feet long and 6 feet wide. How many feet of fencing does Ken need to buy?

   Write your plan here.   **Show your work here.**

Chapter 5

Mighty Castles

Looking Back

To: _____
(Your Name)
From: ivantheinventor@timetravels.org

subject: Time Travel Trip

I have just invented a time travel machine. I am looking for someone to try it out. I know you like adventures. How would you like to travel back in time 700 years to live in a castle in England? Please let me know.

To: ivantheinventor@timetravels.org
From: _____
(Your Name)

subject: *re*: Time Travel Trip

YES! I would like to sign up for this trip. Thank you for thinking of me.

Understand   Plan   Try   ▶ **Look Back**

In this chapter, you will *look back* at castles from long ago. You'll also learn different ways to *look back* at a math word problem to check your work after you've solved it. You will practice using the fourth step of the four-step problem-solving method: **Look Back**.

▲ Between the years 1000 and 1300, many castles were built in England.

▼ A ditch filled with water, called a *moat*, often surrounded many of the castles. People walked over a drawbridge to get in and out of the castle.

▲ There were no refrigerators in the days of castles. Salt was used on meat, fish, and vegetables to preserve them.

▶ Castle musicians played an instrument called the *hurdy-gurdy*.

73

Answering the Question Asked

Always look back to make sure you answered the question.

You want to learn about castles before you go on the time travel trip. You check out books from the library.

You are looking at a picture of Dover Castle. You wonder,

How long did it take to build Dover Castle?

You search through your books to find out. You find the following information.

> Dover Castle is 80 feet high.

The information tells you something about the castle, but it does not answer your question.

> The water well at Dover Castle is 350 feet deep.

The information is interesting, but again it does not answer your question.

> It took 10 years to build Dover Castle.

Finally, you have found the answer to your question.

When you solve a word problem, it is important to always ▶ **Look Back** at the problem to be sure you answered the question asked.

Understand   Plan   Try   ▶ Look Back

You are interested in comparing Dover Castle to other castles.

Castle Heights and Walls

Name	Height	Thickness of Wall
Castle Rising	50 feet	7 feet
Dover Castle	83 feet	12 feet
Kenilworth Castle	80 feet	14 feet

Circle the choice that answers the question. Use the table.

1. Dover Castle and Kenilworth Castle are about the same height. Which castle is higher? ◀ MTK 268, 346

   A. Dover Castle is higher than Kenilworth Castle.

   B. Dover Castle is 83 feet high.

   C. Kenilworth Castle has thicker walls than Dover Castle.

2. Tell why you made the choice you did for exercise 1.

   _____

   _____

3. What is the name of the castle with walls that are twice as thick as the walls of Castle Rising? ◀ MTK 268, 346

   A. The walls of Dover Castle are not as thick as those at Kenilworth Castle.

   B. 14 feet is twice as thick as 7 feet.

   C. Kenilworth Castle has walls that are twice as thick as the walls of Castle Rising.

Huge fireplaces provided the only source of heat. The castle walls were built very thick to help keep the heat in.

4. Tell why you made the choice you did for exercise 3.

   _____

   _____

more ▶

Think about why an answer would not match the question asked. Then, find the correct answer that would.

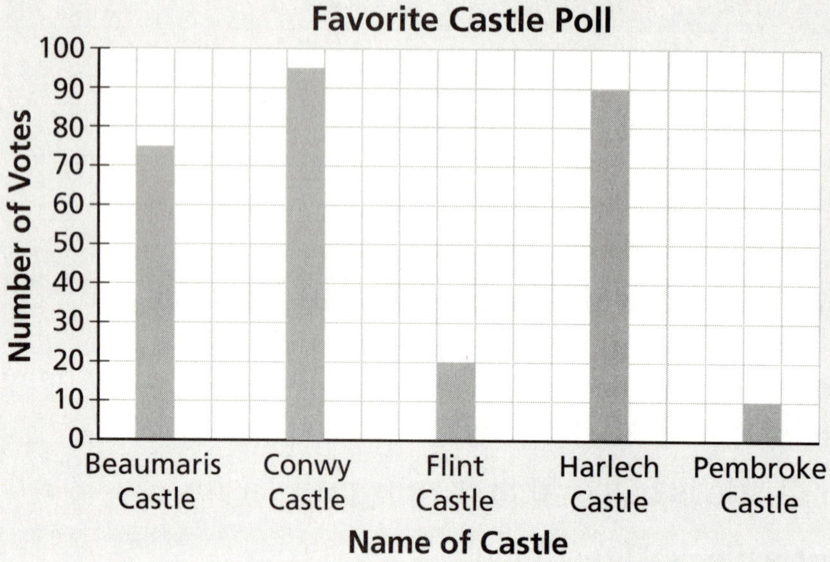
Favorite Castle Poll
Number of Votes
100
90
80
70
60
50
40
30
20
10
0
Beaumaris Castle
Conwy Castle
Flint Castle
Harlech Castle
Pembroke Castle
Name of Castle

Circle the choice that answers the question. Use the graph.

5. Did Conwy Castle or Harlech Castle get more votes? ◂MTK 273

A. Conwy Castle and Harlech Castle got more votes than any other castle.

B. Conwy Castle got more votes than Harlech Castle.

C. The difference between the number of votes for Harlech Castle and for Conwy Castle is 5.

6. Tell why you made the choice you did for exercise 5.

____

____

____

7. Which castle got 9 times as many votes as Pembroke Castle? ◀MTK 273

   A. Harlech Castle got 9 times as many votes as Pembroke Castle.

   B. 90 votes is 9 times as many as 10 votes.

   C. Harlech Castle got 90 votes.

8. Tell why you made the choice you did for exercise 7.

   _____

   _____

   _____

9. How many more votes did Conwy Castle get than Beaumaris Castle? ◀MTK 273

   A. Conwy Castle got 95 votes and Beaumaris Castle got 75 votes.

   B. Conwy Castle got more votes than Beaumaris Castle.

   C. Conwy Castle got 20 more votes than Beaumaris Castle.

10. Tell why you made the choice you did for exercise 9.

    _____

    _____

    _____

It's important to read a word problem again after you're done. Check to be sure you've answered the question.

Comparing Your Answer to Another Number

Compare your answer with other numbers in the word problem. Then, cross out answer choices that are not reasonable.

You learn that drinking water in the castles came from wells. These wells were very deep in the ground.

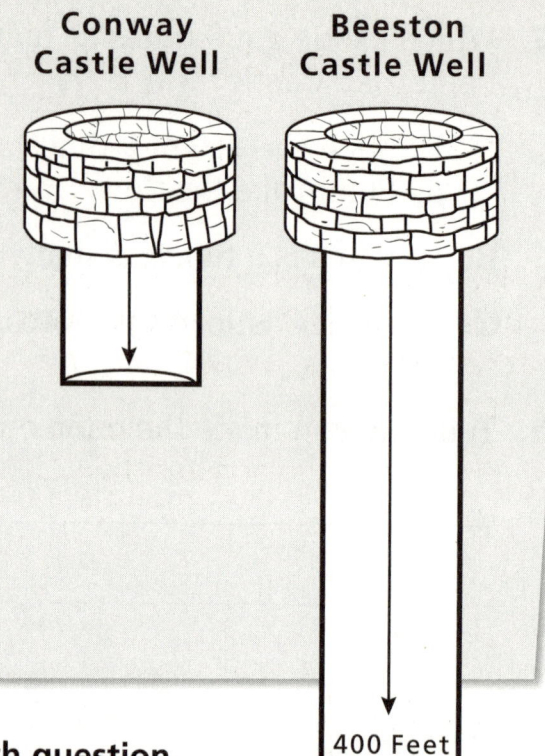

Conway Castle Well

Beeston Castle Well — 400 Feet

Use the drawing to help you answer each question.

1. Is the depth of the Conwy Castle well *more than* or *less than* the depth of the Beeston Castle well? ◀ MTK 12–13, 346

   _____less than_____

2. Is the depth of the Conwy Castle well *more than* or *less than* 400 feet? ◀ MTK 12–13, 346

   _____

3. Would 500 feet be a reasonable estimate of the depth of the Conwy Castle well? ◀ MTK 128–130, 346

   _____

4. Would 100 feet be a reasonable estimate of the depth of the Conwy Castle well? ◀ MTK 128–130, 346

   _____

Vocabulary ▼ reasonable

Understand   Plan   Try   ▶ Look Back

The day has finally arrived for your trip. You enter the time travel machine and *Swoosh!* you are inside a castle in medieval England.

DO NOT solve exercises 5–7. Use the THINK to help you cross out the two answer choices that are not reasonable.

5. You decide to hike along the castle wall. You walk 630 feet to a lookout, then another 320 feet. How far did you walk in all? ◀ MTK 12–13, 146–151, 346

   THINK: Is the answer *more than* or *less than* 630 feet?

   _____more than_____

   A. ~~320 feet~~   B. 920 feet   C. 950 feet   D. ~~600 feet~~

6. You are hungry. The castle cook is baking bread. The recipe calls for 2 cups of flour for each loaf. How many loaves can the cook bake with 18 cups of flour? ◀ MTK 78–79, 356

   THINK: Is the answer *more than* or *less than* 18 loaves?

   _____

   A. 20 loaves   B. 9 loaves   C. 36 loaves   D. 2 loaves

7. This kitchen has its own water well. The cook stores 3 full buckets of water in a large tub. If each bucket holds 4 gallons, at least how many gallons of water are in the tub? ◀ MTK 64–65, 356

   Did you know?
   Three types of bread were baked in medieval times. There was the *manchet* for the lord, *brown cheat* for the servants, and *brom-bread* for the dogs and horses.

   THINK: Is the answer *more than* or *less than* 4 gallons?

   _____

   A. 1 gallon   B. 3 gallons   C. 12 gallons   D. 17 gallons

   more ▶

Vocabulary ▽ not reasonable ▽ cup (c) ▽ gallon (gal)

Ask a question to help you decide if an answer makes sense.

DO NOT solve exercises 8–9. Use the THINK to help you cross out the two answer choices that are not reasonable.

8. A boy is ready to walk on stilts. He is about 4 feet tall. The stilts make him 2 feet taller. About how tall will he be on the stilts? ◀ MTK 12–13, 34–36, 346

   THINK: Is the answer *more than* or *less than* 4 feet?

   _____

   A. 3 feet   **B.** 6 feet   **C.** 8 feet   **D.** 2 feet

   Walking on Stilts

9. Someone at the castle rings the bell every 3 hours to keep track of time. You hear a bell ring when you first start working in the garden. Later, you hear the bell ring two more times. When you hear the bell ring the third time you stop working. How many hours did you work? ◀ MTK 34–36, 334

   THINK: Is the answer *more than* or *less than* 3 hours?

   _____

   A. 6 hours   **B.** 2 hours   **C.** 1 hour   **D.** 9 hours

Understand   Plan   Try   ▶ Look Back

You decide to practice archery. You make a table to keep track of how well you are doing.

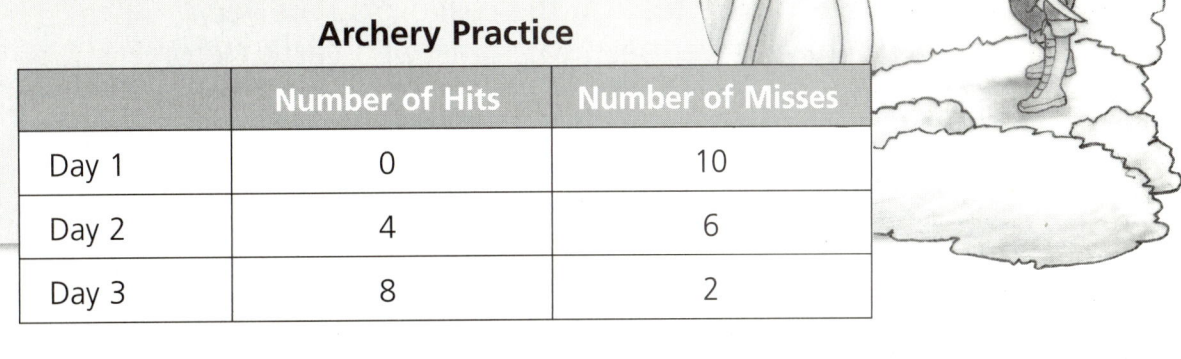

Archery Practice

	Number of Hits	Number of Misses
Day 1	0	10
Day 2	4	6
Day 3	8	2

DO NOT solve exercises 10–11. Look at the table. Use the THINK to help you cross out the two answer choices that are <u>not</u> reasonable.

10. How many arrows did you shoot on Day 2? ◀ MTK 12–13, 34–36, 268

    THINK: Is the answer *more than* or *less than* 6 arrows?

    _____

    A. 12 arrows   **B.** 4 arrows   **C.** 10 arrows   **D.** 2 arrows

11. Suppose you hit the target 2 fewer times on Day 2. How many times would you have hit the target for all 3 days? ◀ MTK 12–13, 34–36, 46–48, 268

    THINK: Is the answer *more than* or *less than* 8 times?

    _____

    A. 10 times   **B.** 4 times   **C.** 6 times   **D.** 12 times

12. In exercise 11, how did you decide whether your answer was *more than* or *less than* 8 times?

    _____

    _____

    _____

81

Using the Correct Label for Your Answer

Always look back to be sure you have correctly labeled your answer.

There is going to be a feast at the castle. You are very excited. You have never been to a feast at a castle before. The cooks ask for your help.

To get the ovens ready, they need you to bring in pieces of wood. You run out of the kitchen just as the cook said,

"Now, be sure each piece is no longer than 3 …"

Uh-oh, did the cook say 3 inches or 3 feet? You wonder.

There is a big difference between *3 inches* and *3 feet*. You need to know the correct length. That is why labels are very important when you answer a math word problem.

Choose the correct label from the box to answer exercises 1–3.

1. There will be 5 tables in the Great Hall that can seat 20 people each and 7 tables that can seat 30 people each. How many people can be seated at the party? ◀ MTK 121, 146–151, 172–175

   | tables |
   | people |
   | chairs |
   | seats |

   Answer 310 _____

82

Understand   Plan   Try   ▶ Look Back

2. You decide to make an outfit to wear to the feast. You need to cut a piece of velvet that is 6 yards long into 3 equal pieces. How long will each piece be? ◀ MTK 78–80, 346

   Answer 2 _____

   | pieces |
   | outfits |
   | yards |

3. A few days earlier, the cooks made two different pies for people to taste. They cut each pie into fourths. How many pieces did they cut? ◀ MTK 210–213

   Answer 8 _____

   | pieces |
   | pies |
   | cooks |

Write the correct label to answer exercises 4–5.

4. One of your jobs is to get the balls for the jugglers. There will be 7 jugglers. Each juggler needs 6 balls. How many balls do you need to get? ◀ MTK 60–62

   Answer 42 _____

5. Another one of your jobs is to hang flags along the sidewalls in the Great Hall. You need to hang 12 flags each on two of the sidewalls. How many flags will you need to hang up? ◀ MTK 146–147, 172–173

   Answer 24 _____

Vocabulary ▪ fourth

Using Estimation to Check Your Answer

Often you have many choices for an answer to a word problem.

Use an estimate to rule out the choices that are <u>not</u> reasonable.

DO NOT solve exercises 1–4. Follow the steps to rule out answer choices that are <u>not</u> reasonable.

1. To decorate the Great Hall you put a ribbon around the four walls of the hall. The hall is 88 feet long and 23 feet wide. How much ribbon do you need? ◀ MTK 132–133, 262, 346

   Circle the choice that gives the best estimate of the answer.

   (**A.** 90 + 20 + 90 + 20)   **B.** 90 + 20   **C.** 90 × 20

   What is your estimate for the answer? About ___220___ feet.

   Use your estimate to cross out two answers that are <u>not</u> reasonable.

   A. ~~11 feet~~   **B.** ~~110 feet~~   **C.** 226 feet   **D.** 222 feet

2. To freshen up the stuffy castle, you make pomanders. The cook has 9 oranges and 567 cloves. You want to use up all the cloves by putting the same number of cloves on each orange. How many cloves would you use for one orange? ◀ MTK 139

   Castles were very smelly places. People used to carry oranges with cloves stuck in them, called *pomanders*—medieval air fresheners!

   Circle the choice that gives the best estimate of the answer.

   A. 570 − 10   **B.** 570 + 10   **C.** 570 ÷ 10

   What is your estimate for the answer? About _____ cloves.

   Use your estimate to cross out two answers that are <u>not</u> reasonable.

   A. 560 cloves   **B.** 57 cloves   **C.** 580 cloves   **D.** 63 cloves

Understand   Plan   Try   ▸ Look Back

3. There are 12 tables set for the feast. Your job is to put 5 candles on each table. How many candles do you need? ◂ MTK 136–137

Circle the choice that gives the best estimate of the answer.

A. 10 − 5   **B.** 10 + 5   **C.** 10 × 5

What is your estimate for the answer? About _____ candles.

Use your estimate to cross out two answers that are not reasonable.

A. 15 candles   **C.** 17 candles

B. 60 candles   **D.** 50 candles

4. A total of 302 people attended this feast. Last year, 293 people came to the feast. How many more people came this year? ◂ MTK 132–133

Circle the choice that gives the best estimate of the answer.

A. 300 − 290   **B.** 300 + 290   **C.** 300 × 290

What is your estimate for the answer? About _____ more people.

Use your estimate to cross out two answers that are not reasonable.

A. 590 more people

B. 10 more people

C. 9 more people

D. 595 more people

more ▸

An estimate can help you rule out unreasonable choices.

The feast is ending. It has been the best feast ever.

DO NOT solve exercises 5–8. Follow the steps to rule out answer choices that are not reasonable.

5. You have 184 pieces of candy in a basket. You decide to give them to the 8 cooks at the castle. If each cook gets the same number of candies, how many pieces of candy will that be? ◂MTK 140

   Circle the choice that gives the best estimate of the answer.

   A. 184 + 10   **B.** 184 − 10   **C.** 180 ÷ 10

   What is your estimate for the answer? About _____ pieces.

   Use your estimate to cross out two answers that are not reasonable.

   A. 194 pieces   **B.** 174 pieces   **C.** 18 pieces   **D.** 23 pieces

It is time to return to your life back home. As a going away gift, the lord of the castle gives you a suit of armor for a horse and for you.

6. Your suit of armor weighs 54 pounds. The one for the horse weighs 106 pounds. How much weight will this add to the time travel machine? ◂MTK 132–133, 358

   Circle the choice that gives the best estimate of the answer.

   A. 100 + 50   **B.** 100 − 50   **C.** 100 ÷ 50

   What is your estimate for the answer? About _____ pounds.

   Use your estimate to cross out two answers that are not reasonable.

   A. 160 pounds   **C.** 50 pounds

   B. 150 pounds   **D.** 2 pounds

Understand   Plan   Try   ▶ Look Back

You step back into the time machine and *Swoosh!* you are back home. You ask Ivan the Inventor if you can use the time machine to send some presents to your friends at the castle.

7. You remember that the people at the castle loved your baseball cap. You can buy a box of 25 caps for $79. You decide to buy 8 boxes. How much did you spend on the caps? ◀ MTK 17, 136–137

Circle the choice that gives the best estimate.

A. 30 + 80 + 10   **B.** 80 × 8   **C.** 80 + 8

What is your estimate for the answer? About $ _____.

Use your estimate to cross out two answers that are not reasonable.

A. $640   **B.** $120   **C.** $88   **D.** $632

8. As a special treat, you send your friends at the castle some jellybeans. At the grocery store you spend $15 for 6 pounds of jellybeans. How much did the jellybeans cost a pound? ◀ MTK 17, 139

Circle the choice that gives the best estimate.

A. 18 × 6   **B.** 18 ÷ 6   **C.** 18 − 6

What is your estimate for the answer? About $ _____.

Use your estimate to cross out two answers that are not reasonable.

A. $10.80   **B.** $2.88   **C.** $3   **D.** $12.80

87

Chapter 5 Test

Fill in the circle with the letter of the correct answer.

1. Carlos measures the length of his desk. Which of the following might be the length?

    (A) 36 hours
    (B) 36 pounds
    (C) 36 inches
    (D) 36 days

2. Jana sold 21 gift-wrap orders one year. The next year she sold 36 orders. Which of the following best describes the number of orders she sold in those two years?

    (A) More than 36 orders
    (B) Exactly 36 orders
    (C) Less than 36 orders
    (D) Less than 2 years

3. Carolina plans to practice the violin 20 minutes a day to get ready for the recital next month. Which unit would you use to describe the amount of time spent after 3 days of practice?

    (A) minutes
    (B) violins
    (C) time
    (D) A.M.

4. Mike had 48 tickets to a baseball game. He sold some of those tickets and has 31 left. How many tickets did he sell?

    (A) 47 tickets
    (B) 37 tickets
    (C) 27 tickets
    (D) 17 tickets

Fill in the circle with the letter of the correct answer. Tell why you made your choice.

5. Ngyuen wants to put a wallpaper border around his room. His room is 20 feet long and 15 feet wide. How long does the wallpaper border need to be?

   (A) 35 feet _____

   (B) 40 feet _____

   (C) 70 feet _____

   (D) 90 feet _____

For problem 6, write your plan and show your work.

6. Anna is running for class president. She has 56 *Vote For Anna* buttons. She gives the same number of buttons to 7 friends to help her pass them out at lunchtime. How many buttons does each friend get?

   Show your work here.

   _____

   _____

   _____

7. Write two ways that you can look back to check the answer to a word problem.

   1) _____

   2) _____

89

Chapter 6

Ribbit, Ribbit
Putting It All Together

Welcome to: Fantastic Frogs.com

- Frog Stories
- See Live Frogs
- Frog Toys
- Frog Facts

Ask:

Hello and welcome to my website. I'm Freddy the Frog. If you have a question about frogs, just ask me. I'm an expert—your source for anything you want to know about frogs.

HELP WANTED

Freddy the Frog needs a helper to answer questions about frogs. Must know a lot about frogs and have good problem solving-skills.

▶ **Understand**   ▶ **Plan**   ▶ **Try**   ▶ **Look Back**

In this chapter, you will see that an organized method can help you answer all those questions about frogs. You'll put together all the skills you've learned. You'll see how the four-step problem-solving method (**Understand, Plan, Try,** and **Look Back**) can help you work with Freddy and become an expert problem solver at the same time.

Frogs have been on Earth for over 190 million years.

◀ In Japan, frogs are symbols for good luck.

▶ Some frogs can blend in with their environment so other animals can't find them. For example, the Darwin's frog can float upside down in the water. This way, it looks like a fallen leaf instead.

▲ Fossil of a frog.

Many languages have a special sound that identifies a frog. Here are some examples:

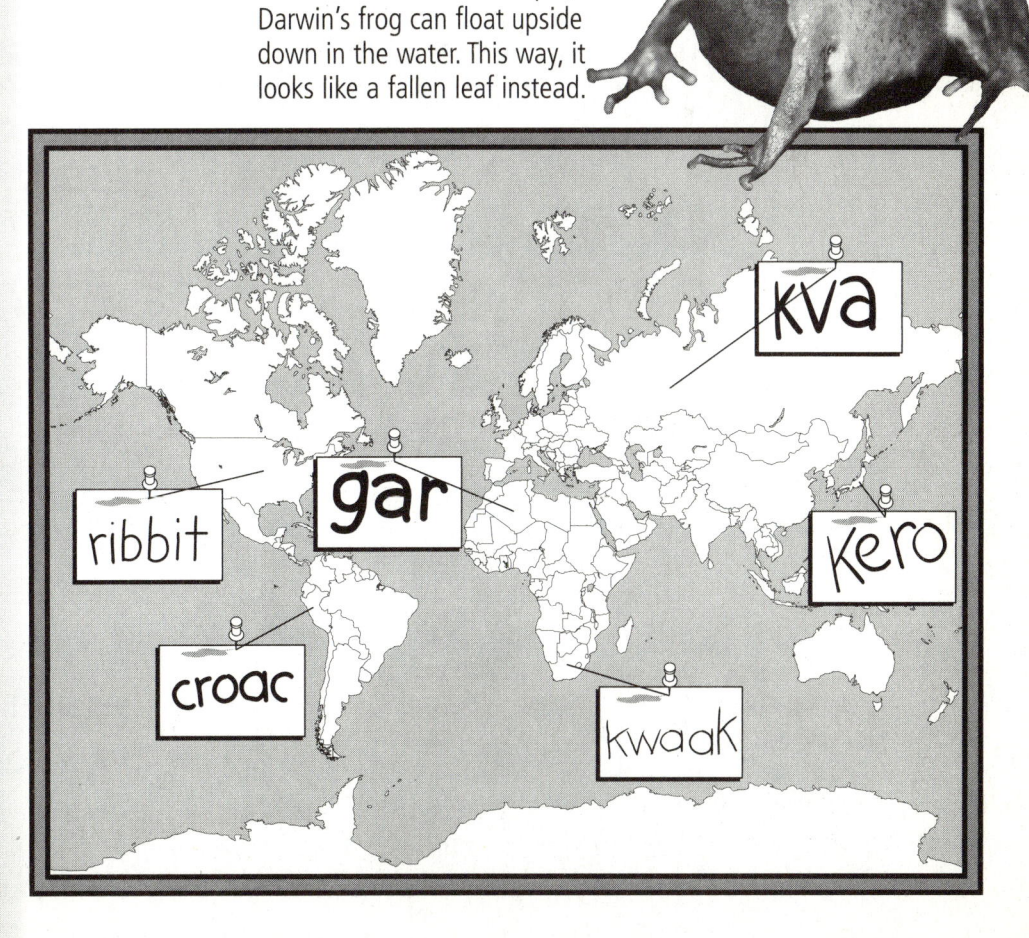

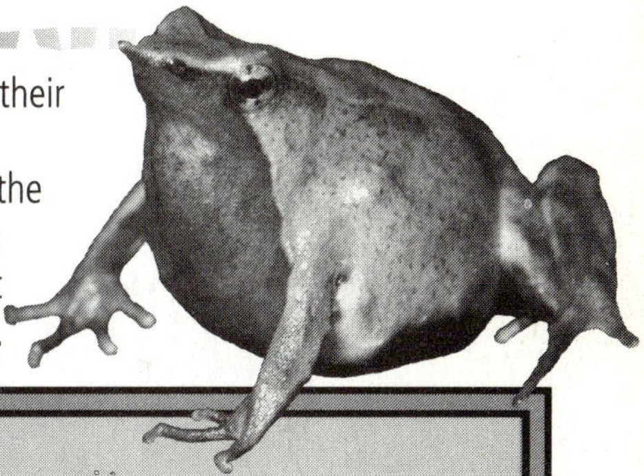

kva

gar

ribbit

kero

croac

kwaak

91

Using the Four-Step Problem-Solving Method

Now it's time to apply what you've learned.

You really want to work with Freddy the Frog. You meet with him to find out more about the job.

> Here is what I do when a letter comes in to the website. First, I need to **understand** the question. If there is a problem to solve, I make a **plan** to solve it. Then I **try** out my plan. And, finally, before I respond, I **look back** to make sure I've answered the question.

▸ **Understand**   ▸ **Plan**   ▸ **Try**   ▸ **Look Back**

Freddy is one smart frog. What he does to answer questions sounds a lot like what you do in school to solve math word problems.

You tell Freddy about the four-step problem-solving method.

FOUR-STEP PROBLEM-SOLVING METHOD

Step 1 ▸ **Understand** the problem.

Step 2 ▸ **Plan** how to solve the problem.

Step 3 ▸ **Try** your plan.

Step 4 ▸ **Look Back** at your solution to check it.

Freddy is very interested in this. He asks you for more details.

Name the step or steps you learned in each chapter.

Chapter 1 _____

Chapter 2 _____

Chapter 3 _____

Chapter 4 _____

Chapter 5 _____

Freddy is very impressed. He gives you the job.

Now you can put all four steps together to help Freddy solve the problems that are sent to the website.

Sometimes a simple sketch can help you decide how to solve a word problem.
Problem

Dear Freddy,

Here is a problem about the Malay gliding frog. Suppose there are 4 trees in a straight line in the rainforest. The trees are 32 feet apart from each other. If a gliding frog starts at the first tree and glides from tree to tree until it gets to the last tree, about how far will it have traveled?

Joaquin

The Malay gliding frog fills the webbing between its toes with air so that it can glide from tree to tree. It can glide up to 50 feet.

Answer the questions to see how the four-step problem-solving method can be used to solve the word problem.

▸ **Understand**

1. What does the problem ask you to find?

   _____

   _____

2. Place a ✔ in the ☐ if that information is needed to solve the problem.
   ☐ There are 4 trees in a straight line.
   ☐ The Malay gliding frog glides with its feet.
   ☐ The trees are 32 feet apart.

▸ **Plan**

3. Circle the sketch that can help you find how far the frog glides.

   A.
   32 feet 32 feet 32 feet

   B.
   32 feet 32 feet 32 feet 32 feet

94

▸Understand ▸Plan ▸Try ▸Look Back

4. Circle the plan you can use to solve the problem. ◀ MTK 146, 346

   Plan A
   - Multiply 32 by 4.
   - The answer will be in feet.

   Plan B
   - Multiply 32 by 3.
   - The answer will be in feet.

   Plan C
   - Multiply 32 by 5.
   - Write *feet* after 58 for a label.

▸**Try**

5. Show how you would carry out the plan. ◀ MTK 172–173, 346

6. Write a sentence that answers the problem.

   _____

▸**Look Back**

7. Did you answer the question that was asked? _____
   If you answered *no*, go back and redo your work.

8. Circle a good estimate for the answer. ◀ MTK 136–137

   A. 30×3　　　**B.** 30×4　　　**C.** 30×5

9. Does your estimate show that your answer is reasonable? _____
   If you answered *no*, go back and check your work.

10. Does your answer have the correct label? _____
    If you answered *no*, go back and add or change the label.

more ▸

Sometimes you can use a simpler problem
to help plan your solution.

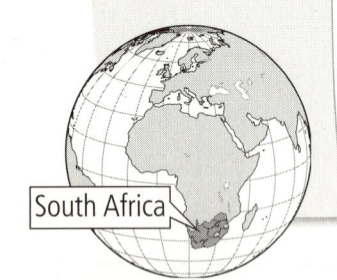

Hi, Freddy,

A frog named Burning Brightly jumped about 19 feet in a contest. I read that a frog named Santjie jumped even longer than that at a derby in South Africa. How close did Burning Brightly come to tying Santjie's record? Thanks!

Tamara

Answer the questions to see how the four-step problem-solving method can be used to solve the problem.

▶ **Understand**

11. What does the problem ask you to find?

Did you know? Santjie jumped about 33 feet at the derby.

_____

12. Place a ✔ in the ☐ for the information you need to solve the problem.

   ☐ Burning Brightly's jump was about 19 feet long.

   ☐ Santjie's jump was about 33 feet long.

   ☐ Santjie jumped at a frog derby in South Africa.

▶ **Plan**

13. Use this simpler problem to help you plan what to do.

Think! **Simpler problem** Green frog jumps 2 feet. Yellow frog jumps 6 feet. How much longer is the yellow frog's jump?

Draw a sketch that describes the simpler problem.

Circle the choice that describes the simpler problem. ◀ MTK 251, 346

A. 6 ÷ 2   B. 6 × 2   C. 6 + 2   D. 6 − 2

Circle the choice that describes the original problem.

A. 33 ÷ 19   B. 33 × 19   C. 33 + 19   D. 33 − 19

▸ **Understand** ▸ **Plan** ▸ **Try** ▸ **Look Back**

▸Try

14. Show how to carry out the plan. ◂MTK 161

15. Write a sentence that answers the problem.

_____

_____

Look Back

16. Did you answer the question that was asked? _____

If you answered *no*, go back and redo your work.

17. Circle a good estimate for the answer. ◂MTK 132–133

    A. 30 − 20 = 10   **B.** 3 − 2 = 1   **C.** 30

18. Does your estimate show that your answer is reasonable?

_____

If you answered *no*, go back and check your work.

19. Does your answer have the correct label?

_____

If you answered *no*, go back and add or change the label.

more ▶

97

Sometimes the information you need is in a table.

Goliath frog

Poison Dart frog

Hey, Freddy!

I've heard that the Goliath frog is the biggest frog in the world. How many inches longer is the Goliath frog than the Poison Dart frog?

Mike

Length of Frogs

Kind of Frog	Length
Bullfrog	8 inches
Goliath frog	12 inches
Leopard frog	5 inches
Poison Dart frog	2 inches

Answer the questions to see how the four-step problem-solving method can be used to solve the problem.

▸ **Understand**

20. What does the problem ask you to find?

_____

21. Which numbers in the table will you use? Circle them. ◂ MTK 268

22. Is there any other information you need to solve the problem?

_____

If you answered *yes*, how can you find that information?

_____

98

> Understand   > Plan   > Try   > Look Back

> **Plan**

23. Complete the plan. Fill in each blank with a word from the box. ◄ MTK 48, 346

| add |
| subtract |
| multiply |
| divide |
| feet |
| inches |
| from |
| by |

Plan

_____ 2 from 12.

The answer will have the label _____.

> **Try**

24. Show how to carry out the plan. ◄ MTK 48

Did you know?
The Poison Dart frog is so poisonous that it doesn't have to hide. It comes in bright colors like yellow, red, and blue. These bright colors warn other animals to stay away.

25. Write a sentence that answers the problem.

_____

_____

_____

> **Look Back**

26. Did you answer the question that was asked? _____
If you answered *no*, go back and redo your work.

27. Do you think your answer should be *more than* or *less than* 12 inches? _____ Is it? _____
If you answered *no*, go back and check your work.

28. Does your answer have the correct label? _____
If you answered *no*, go back and add or change the label.

more ▶

99

Sometimes a word problem has more information than you need.

Dear Freddy,

I read that in 1960 there were 23 different kinds of Glass frogs known. Since that time, 87 more have been discovered. At least 74 of these Glass frogs live in Colombia, South America. How many different kinds of Glass frogs are there now?

Jennie

Did you know?
A Glass frog has clear, see-through skin. You can watch its heart beat and see what it has eaten.

Answer the questions to see how the four-step problem-solving method can be used to solve the problem.

▸ **Understand**

29. What does the problem ask you to find?

_____

30. What information do you need to solve it?

_____

_____

31. Is there any information that you don't need to solve the problem? If so, you should cross it out in the word problem.

▸ **Plan**

32. Write a plan for solving the problem. ◂ MTK 148

    Plan

    _____

    _____

100

▶ **Understand** ▶ **Plan** ▶ **Try** ▶ **Look Back**

▶ **Try**

33. Show how you would carry out the plan. ◂ MTK 148

34. Write a sentence that answers the problem.

_____

_____

▶ **Look Back**

35. Did you answer the question that was asked? _____

If you answered *no,* go back and redo your work.

36. Do you think your answer should be *more than* or *fewer than*

87 glass frogs? _____ Is it? _____

If you answered *no,* go back and check your work.

37. Explain how you decided whether your answer should be *more than* or *fewer than* 87 Glass frogs.

_____

_____

_____

38. Does your answer have the correct label?

_____

If you answered *no,* go back and add or change the label.

Solving Math Word Problems on Your Own

When solving word problems on your own, you can always use the four-step method.

It won't solve the word problem for you, but it can help you keep organized. Here are some questions to think about as you use the four-step method.

▶ Understand

- What does the word problem ask me to find?
- Do I know what each word in the problem means? (If you don't know what a word means, use your handbook *Math to Know*, your math book, or a dictionary to help you.)
- Should my answer be an estimate or an exact number?
- Is there any information that is missing? If there is, where can I find that information?
- Is there any information that I don't need?

▶ Plan

- Can I draw a sketch to help me solve the word problem?
- Can a simpler similar problem help me solve the original problem?
- Do I need more than one step to solve the word problem?

▶ Try

- Am I carrying out each step of my plan?
- Am I using the correct numbers from the word problem?

▶ Look Back

- Did I answer the question asked?
- Did I label my answer correctly?
- Can I use estimation to check whether my answer is reasonable?

▸ **Understand**　▸ **Plan**　▸ **Try**　▸ **Look Back**

Solve the problem. Explain how you used each of the four problem-solving steps. Use the questions on page 102 to help you.

1. ◂ MTK 161, 360

   > Hi, Freddy!
   > I read in a science book that Wood frogs can live at a temperature of 19° Fahrenheit. How much colder is this than the freezing point of water?
   > Laverne

   ▸ **Understand**

   _____
   _____

   ▸ **Plan**

   _____
   _____

   ▸ **Try**

   Did you know?
   Water freezes at 32° Fahrenheit.

   Answer in a complete sentence.

   _____
   _____

   ▸ **Look Back**

   _____
   _____
   _____

more ▸

Use everything you have learned so far to solve the word problems.

Did you know?
Frog eggs hatch into tadpoles. Wood frog tadpoles change into frogs in about 63 days.

Solve the word problem. Explain how you used each of the four problem-solving steps. Use the questions on page 102 to help you.

2. Hi, Freddy,

   Here is another question. We have some Wood frog tadpoles. They hatched on April 1. Will they change into frogs by my birthday on June 20?

   Laverne

▸ **Understand**

_____

▸ **Plan**

_____

_____

▸ **Try** ◂ MTK 146–147, 342–343

Answer in a complete sentence.

_____

▸ **Look Back**

_____

_____

▸ Understand ▸ Plan ▸ Try ▸ Look Back

3.

Hiya, Freddy,

I hear that you have T-shirts for sale. I'd like to buy 3 children's T-shirts and 2 adult T-shirts. How much will this cost all together?

Freddy

P.S. Did you notice that we have the same name?

▸ **Understand**

_____

_____

Freddy the Frog T-Shirts

Size	Price
Child	$15
Adult	$20

▸ **Plan**

_____

_____

▸ **Try** ◂ MTK 17, 146–147, 174–175

Answer in a complete sentence.

_____

▸ **Look Back**

_____

_____

Now you know how to use the four-step problem-solving method. You can use it whenever you solve math word problems. You may want to keep this book handy, so that you can use the questions on pages 102 to help you.

Good luck and have fun solving word problems!

Chapter 6 Test

Questions 1–5 are about this word problem:

Janine went to the store. She saw a sign saying the red buttons cost 10¢ each, the blue buttons cost 16¢ each, and the yellow buttons cost 19¢ each. She bought 1 red button and 3 yellow buttons. How much money did she spend in all?

For questions 1–2, fill in the circle with the letter of the correct answer.

1. Which information is *not* needed to solve the problem?

   (A) Red buttons cost 10¢ each.

   (B) Blue buttons cost 16¢ each.

   (C) Yellow buttons cost 19¢ each.

2. Which plan would you use to solve the word problem?

   (A) **Plan A**
   - Add the cost of 1 yellow button to the cost of 1 red button.
   - Add the cost of the 2 buttons to the cost of 1 red button.

   (B) **Plan B**
   - Multiply the cost of 1 yellow button by 3 to find the cost of 3 yellow buttons.
   - Add the cost of 3 yellow buttons to the cost of 1 red button.

   (C) **Plan C**
   - Multiply the cost of 1 yellow button by 3 to find the cost of 3 yellow buttons.
   - Add the cost of 3 yellow buttons, the cost of 1 blue button, and the cost of 1 red button.

▶ **Understand**   ▶ **Plan**   ▶ **Try**   ▶ **Look Back**

For questions 3–6, write your answer in the space provided.

3. Carry out the plan from question 2.

4. Write a sentence that gives the correct answer to the word problem.
   _____

5. Explain how estimation can help you check whether your answer in question 4 is reasonable.
   _____
   _____
   _____

6. What are the problem-solving steps you learned in this book?
   _____
   _____

   Name 3 reasons why you think the steps are helpful.

   1) _____
   _____

   2) _____
   _____

   3) _____
   _____

Vocabulary

A A.M.

add

at least

average

B bar graph

108

Vocabulary

billion

cup (c)

cent (¢)

data

circle

109

Vocabulary

degrees (°)

difference

denominator

dime

divide

110

Vocabulary

dollar ($)

even

E **equal**

exact

estimate

F **Fahrenheit (F)**

Vocabulary

feet (ft)

first

fifth

fraction

Vocabulary

G **gallon (gal)**

greater than (>)

H **half**

graph

113

Vocabulary

hexagon

hour (h)

I **inch (in.)**

K **kilogram (kg)**

L **left**

Vocabulary

less than (<)

mile (mi)

long

million

M meter (m)

minute (min)

115

Vocabulary

more than (>)

not reasonable

multiply

number line

nickel

Vocabulary

number sentence

one third

numerator

Vocabulary

P.M.

pictograph

plus

pound (lb)

quarter

118

Vocabulary

R reasonable

rectangular

rectangle

S second (adj)

second (sec)

119

Vocabulary

side

square

skip count

solution

subtract

Vocabulary

sum

tally mark

survey

table

temperature

Vocabulary

third

total

triangle

ton (t)

Vocabulary

twice

wide

U under

Y yard (yd)

W weight

year

123